Marion Dillon

HUMAN DEVELOPMENT

HUMAN DEVELOPMENT

An introduction to the psychodynamics of growth, maturity and ageing

THIRD EDITION

Eric Rayner

London
UNWIN HYMAN
Boston   Sydney   Wellington

Published by the Academic Division of

Unwin Hyman Ltd
15/17 Broadwick Street, London W1V 1FP

Allen & Unwin Inc.,
8 Winchester Place, Winchester, Mass. 01890, USA

Allen & Unwin (Australia) Ltd,
8 Napier Street, North Sydney, NSW 2060, Australia

Allen & Unwin (New Zealand) Ltd in association with
the Port Nicholson Press Ltd,
60 Cambridge Terrace, Wellington, New Zealand

First published in 1971
Third edition published in 1986
Third impression 1990

British Library Cataloguing in Publication Data

Rayner, Eric
    Human development: an introduction to the
    psychodynamics of growth, maturity and ageing.
    —3rd ed.
    1. Genetic psychology
    I. Title
    155   BF701
    ISBN 0-04-155010-2

Library of Congress Cataloging-in-Publication Data
Rayner, Eric.
    Human development.
    (National Institute social services library; 22)
    Filmography: p.
    Bibliography: p.
    Includes index.
    1. Developmental psychology. I. Title. II. Series:
National Institute social services library; no. 22.
BG713.E38   1986   155   85-23022
ISBN 0-04-155010-2 (pbk.: alk. paper)

Set in 10 on 12 point Sabon by Columns of Reading
and printed in Great Britain by Billings and Sons Ltd,
London and Worcester

Contents

Acknowledgements

It is nearly twenty years since I began to think about writing the first edition of this book. Since then so many people have helped and taught me that I cannot do them individual justice. The list of their names would be so long as to be meaningless. Perhaps my patients have been my first informants; I rarely thank them, but here I can express my debt and gratitude. There have been dozens of students, as well as teachers, from Croydon Technical College, London School of Economics, Brunel University as well as the Institute of Psychoanalysis, who have helped me distil what is written here. Many of the examples are quoted from their words.

Florence Mitchell and Mary Barker must be mentioned by name for it was they who gave me the idea and made possible the writing of the book in the first place.

The actual writing of the First edition was really a joint venture over several years with my first wife Mary Rayner, very much of that work still remains. What is new in this third edition comes largely from the thoughtful expertise of my wife now, Dilys Daws. I also owe a special debt to Joan Willans; she, in typing and correcting the manuscripts has been through the birth of all three editions.

The early editions were inspired most intimately by what my children Sarah, Will and Ben taught me. Now, grown up, they are still in the back of my mind, now my severe but kindly judges. For all these things I thank them.

Preamble

Throughout the book, for the sake of simplicity, a single person, child, or adult will be referred to as 'he' or 'him' no matter which sex they are. I apologize for continuing this traditional usage but no–one has yet invented a pronoun, in English at least, that refers to a person of either sex equally. To replace 'he' by 'she' would also be equally discriminatory. Likewise when men and women are referred to collectively they will be called 'man' or 'men'.

Furthermore, since the legalization of partnership in marriage is no longer sacred for many people, the terms male partner and female partner should be used instead of husband and wife. However this terminology is clumsy and long winded. So 'husband' and 'wife' will often be used for partners in stable, long term relationships whether legally tied or not.

CHAPTER 1

Introduction

Background

This book is written in the belief, which is so obvious as to seem trite, that, among other things, we humans ought to strive not only to love but also to understand each other.

The history of our race is a saga of the drive to master fearful imponderables in our world. It is a chronicle not only of triumphs of enlightenment but also of greed, prejudice, cruelty and pain, albeit warmed at least by glimmerings of sympathy and kindliness. The last three hundred years have seen the emergence of the power of scientific logic and knowledge in the mastery of our physical world. This has proved to be dramatically effective in alleviating some of the worst ills of our forbears. For instance, it has only been by scientific care and attention to complex details that, say, the killer smallpox has been made to disappear from the planet. At the same time the scientist's gift of swift communication has provided the opportunity to broaden awareness and sympathy between people and cultures. With this conjunction of wider sympathy and dispassionate scientific honesty there might be a chance, the nuclear arms race notwithstanding, of a climate of human understanding that could dissolve some of the excesses of prejudice, cruelty and pain of life on earth.

But this combination of science and sympathy is complex and difficult to comprehend, it is also slow to take effect. It is thus frightening to those who greedily desire quick and simple solutions. Such people are prone to gather into tyrannical movements, dreaming of magical remedies for our collective ills. They are prone to obliterate learning which is too complex for them. In many countries seats of scholarship and thought can be physically destroyed. In Britain this obliteration is at present not so obviously violent. But, in any country, because understanding seems to provide no immediate economic benefits, it can easily be

1

starved to death by those who measure the value of things by financial profit only.

Even so, the development of sympathetic and scientific awareness seems most precious for our long term future; it is always worth striving for and protecting.

Sympathy and Science

The systematic intention to cultivate awareness of self and others is fairly new in history. Care and consideration for others has no doubt been present since the dawn of animal group living, more recently it has certainly been strongly codified in religious morality. But the idea that we *ought to be informed* about the workings and needs of our own and the minds of others, before deciding upon actions has only gathered coherence in the last century or less.

This has brought a new dimension to morality. If, for instance, we are concerned with children then it is still necessary to heed the command 'love little children', but this simple idea is no longer enough. With our present knowledge we know we will be of little use to a child unless we are aware of his particular developmental problems and needs. And the same applies to people later in life. This requires a great deal of intellectual and emotional work.

This book is concerned with one aspect of the search for greater knowledge about human beings. It aims to be a primer for those who are setting out to inform themselves systematically about the *emotional vicissitudes* of people's lives. It addresses students of the helping professions, but many other readers might also find it interesting and useful.

The combining, in thought, of scientific objectivity with sympathy has already been stressed as of the greatest value to us. It is, I hope, a keynote of the book. There is no need here to describe scientific thought except, perhaps, to say it is a dispassionate holding together of information about the physical world by means of consistent logic.

The other arm of our body of human knowledge, sympathy, is more elusive. By it is meant the mental act of *identifying with* another person, it occurs when we put ourselves 'in his shoes'. For example when we wince at the sight of someone cutting

2

themselves or find ourselves weeping for their tragedy. At such moments we are *being the other person*. It is not ordinarily logical, but to live happily with others we must use it a very great deal.

Such sympathy is useless in a crude and uncontrolled form. It needs to be blended with objectivity to be harnessed into usefulness. This necessary objectivity is close to being scientific.

Thus the argument that lies behind this book is that sympathetic sensitivity and systematically tested scientific knowledge are together necessary but are not by themselves sufficient to create a climate of sympathetic awareness of other people. In building up such understanding, each one of us needs to use the knowledge gained by others before us. But we also have to discover everything anew for ourselves and practise it. Only through practice can we really believe what we feel and know. What is more, we must not forget that we are so prone to self-deception that we need to be critically aware of our tricks of thought, also our effects on others, if a stable climate of understanding is to be cultivated.

Sympathetic awareness can of course be applied either over a wide abstract vista in comprehending other cultures, or in much narrower focus when making sense of another family's life. Most intimately it is applied when understanding the predicaments of one other person. This last is the concern of the present book.

There are many ways of exercising such sympathetic understanding. It has been done for centuries in ordinary listening and talking, also in the confessional, in cures of souls and the casting out of devils. But, starting with the hypnotists in France two hundred years ago, more ambitious, systematic and committed attempts to help others, by understanding, have gathered force in the movements of modern psychotherapy. Recent decades have seen a multitude of innovations originating in this; counselling, group therapy, encounter groups, sensitivity training, family therapy and others (Ellenburger, 1970).

This book does not consider such direct encounters; it is complementary to them but stands at a slightly greater distance, and is concerned with some of the predominant experiences of people as they grow up, age and die. It will focus upon the intimate, personal or inner thoughts, feelings and actions of individuals or 'single-selves' and how they may structure their private worlds as they go through life.

You may wonder why you should be expected to appreciate a single individual's experiences. There are so many immediate problems of vital importance such as health, poverty, housing or race intolerance, which require a wider vision than understanding individual experiences, that surely these must come first? But we are now in a culture which is in revolution or at least in rapid flux. Nearly every traditional way is being questioned as to whether it is inadequate or even destructive of happy living. As ordinary citizens we cannot help participating, if only by default, in the creation of changed systems of culture. And the recipients, or victims of these cultures will be individual human beings. If we do not deeply study the needs of individuals and attune our cultural practices to them people are likely to become expendable items in new social dreams. Only individuals experience suffering and joy, never social systems.

Biological Evolution

The understanding of individual experiences may be clearer if the stage is set by a brief consideration of biological evolution. Looking at the world, we see that a characteristic of all living things is that they are self-perpetuating organisms. Digesting from its environment and excreting into it, each organism is continuously changing its constituents but keeps its own form. It has a characteristic spirit of its own until it dies. With evolution, the breadth and variety of some organisms' exchanges with the environment have developed enormously. Vertebrates, for instance, can hunt and escape in ways that are impossible for the protozoa. This is particularly so with mammals; a combination of large brains and prolonged infancy has enabled them to develop a range of activity which far exceeds lower animals. Their brains, capable of swift and long–standing learning, are *central control systems* which allow innate reflex patterns to be modified and combined in nigh infinite ways, yet still have coherence for the animal as a whole.

The primates function at a higher level than any other mammal with these characteristics of central control; they can also, unlike most other animals, be sexually aroused almost continuously. A million or so years ago the human seems to have been a

4

meat–eating, group–living, erect, large–brained primate. Thus he must have been a hunter, intelligent about the environment, guileful with prey and wary of threats.

As he lived in a group, the human must have developed sensitivity to and co–operativeness with his fellows. The fittest who survived must have been those who could submit to group decisions and also control their own sexuality as much as those who were masters over them (Aitchison, 1983; Becker, 1971).

Man's brain, with its vast memory and acting as a central control system, makes *thought* possible.

Consider this hypothetical example of a process of thought.

One of our forebears out hunting one day looks at the tracks of animals going down to a watering place by the river. Among many hoof and paw prints he sees some exceptionally large ones of different sizes, he roughly counts and measures their distances apart. His body shivers and he thinks something like this: 'Mammoths, as many as the fingers of both my hands, old and young. One would be food enough for many days. But I cannot do it alone, and mammoths kill. I must warn the women and children and find the hunting men.'

We can use this to ponder for a moment on some characteristics of human thought anywhere. First the man's *body* is involved. Not only is much of his thought a representation of physical actions including those of his own body, but he is *emotional*. He shivers, presumably in excitement, in anticipation of the hunt and food. He is in fear of death and injury for himself and his group. These emotions are body activities on the one hand and ideas on the other.

Next, the man *imagines*, he makes *mental representations* of events, these may conform to external reality or be fantasic, or be a mixture. The representations take place in an underlying *dynamic* setting. That is *end states are being striven for*. In this case our early man desires food and to protect himself and his kin from injury and death.

We also see that his mental representations discriminate and put together imagined *relationships* between things. He is intuitively quite a mathematician. Thus he relates the size of footprints to other prints, he conceives of the relative sizes of animals' feet and of mammoths to himself, and he thinks of the mammoth herd relative to his kin.

5

Furthermore, in his thinking he is continuously and instantaneously *classifying*. This is an essential survival mechanism. He classifies the large prints as being different from the others. He detects them as being *signs* of the class 'mammoth'. He is, albeit primitively, conceptualizing. He dwells on some of the characteristics or subclasses of mammoths: they are large, eatable, killers, etc.

It is to be noted that this classificatory activity involves a continuous *registration of sameness* together with *discrimination of differences*. Thus mammoths are recognized as the same in being large, eatable and dangerous, but different, not only from other animals, but also amongst themselves in size and age.

These are, of course, not by any means an exhaustive list of the characteristics of thought, but they will do to start with, we will be using them later in the book.

The capacity to imagine and conceptualize means that a man can create mental representations of events and make *choices* between alternatives before taking action. If he is sane, he will know the boundaries of his own body and self. He can thus discriminate his self–created, inner or private world of imagination and feeling as distinct from the external world of things and space. Being able to imagine also means that he can delay and conceive of time, the beginning, middle and end of things. He can imagine his own future death.

Being imaginative with a sense of inner and outer reality, of space and of time, he can be reflective and aware of his *self* in a world of other selves. He is capable of wonderment. But there is a price to pay for this wealth of experience; he *must conform* to survive. He is thus at the mercy of others and their madness and limitations as well as finding comfort and enjoyment with them. In conforming it is likely that a human will lose some of his potentialities, often to a crippling degree, and as conformity starts at a very early age the experiences of limitation can be very profoundly felt. Yet without conformity a human can have no order or security.

In being imaginative and needing to be dependent upon and conforming to others, the human experiences threats as well as delights from self and others. He can be aware of *helplessness* especially when young. At the same time he can imagine his *effects on others*. He can thus be aware of, or imagine, his *value* and *guilt*.

To summarize, the human is a large–brained, group–living primate capable of being sexually aroused at almost any time. He is capable of creating mental representations of his world and of choice in using these representations. He can experience helplessness, wonderment, value and guilt.

Attending To an Individual's Activity

Our biological summary has outlined what humans in general are capable of. It is quite different to recognize what a particular person, from a specific background, at a certain point in his life, actually does experience and do. Research has made it plain that any one person's vitality of thought, feeling and action is delimited by a vast web of influences: by the coding of his genes, by his uterine environment, by the bias of the society of his early family which itself is delimited by its culture. Then a person is delimited by the pattern of relationships between his family and those around, his friends and school. Later he will be influenced by work, friends and the new families he encounters as he marries, has children and ages. But this is not all: each person's actions affect those around him and they in turn affect him. A person grows and lives in a living web of circular *feedback* relationships (Konner, 1982; Wagner, 1982).

Every person has his own delimited capacities for feeling and action. The meaningfulness of the world, and our usefulness to it, is different for each one of us. Every person is idiosyncratic. When the question of enhancing the individual person arises, counsellors, teachers, social workers and therapists must come forward and those skilled in abstract generalization – evolutionary biologists, philosophers and academic scientists – must step into the background.

This is the viewpoint of our book. I am a specialized worker with individuals, a psychologist and psychoanalyst by profession, and my underlying argument will be a psychoanalytic one though the observations of many other disciplines will be drawn upon. That is, historically the book is largely based upon the *intimate, detailed work with individual people* originated by Freud nearly a century ago (Freud, 1915b). Since then psychoanalytic ideas have been modified and enlarged throughout the world especially in America, France, and Britain.

7

We need not bother with the details of such theories at this point. It is sufficient to say that we are concerned with a *relaxed way of attending*, with as few preconceptions as possible, to what a person is doing and thinking.

Let us imagine an everyday example of such relaxed attention. We will pretend we are having a cup of tea at the home of an old friend who is a young mother.

As she is cutting up some food and putting it into a pan, she comments, 'Must get this done before Jo wakes up, it won't be done in time if I have to go off to collect the others from school first, and Bill is always so hungry when he gets home'. The baby stirs upstairs, the young woman looks drawn and purse's her lips. She says nothing, then the cloud seems to lift from her face and she says, 'Do you remember that first evening of the sixthform play, oh that applause, I'll never forget it, we were somebody then. But that old cow of a drama teacher: all she would do was tell us where we had fluffed our lines'.

You will note that we have dwelt upon a dispassionate observation of her activity, her movements about the kitchen, the expressions on her face and so on. We have also used sympathy. We felt along with her hurry, her weariness, her sense of duty and her anger. We *identified* ourselves with her. We also noted her shifting away from the present to an enjoyable memory of the past.

We might go on to wonder to ourselves about whether her gloom was transient or whether she was getting more chronically depressed. We could note that her memory of the past was a rest from, even flight from, her present exhaustion and frustration. We may hazard a provisional guess that she doesn't feel she is anybody of note at present and starves for some public acclaim again.

On the basis of this we might say sympathetically, 'Must be rotten feeling you're nothing but a slave to them all'. She might reply, 'Oh yes, but there's nothing like seeing them growing and being happy and Bill's a dear really'. In which case we would perhaps sigh with relief. On the other hand she could say, 'Oh no, the children are such good darlings and Bill is a wonderful man'. In this case we might wonder whether she was really depressed and angry but *denying* it, because she seemed to be overdoing her adoration of the family.

8

In the first case we could be reassured that she was pretty well aware of, and coping with, the complex and conflicting desires and frustrations of her present life. But if the second case happened we might bother about the denial of her angry depressed and dissatisfied feelings. This would mean that she, at the present time at least, is *unaware* or *unconscious* of important emotional aspects of her life.

It is this dwelling upon the emotions of individual people, be they consciously felt, dimly perceived or quite unconscious that has been the hallmark of the psychotherapeutic disciplines. These have perhaps found their most systematic expression in Freud and psychoanalytic thinkers after him.

We will be drawing upon psychoanalytic ideas a great deal but not exclusively so. We have already noted how thought and feeling is rooted in body functions, so the book must rest upon human physiology and anatomy. You should have an elementary acquaintance with them (Beckett, 1918; Miller 1983; Smith 1970, 1984). We are concerned with thought and its growth; research psychologists have contributed a great deal to this subject. They will be referred to repeatedly. Furthermore an individual and his predicaments can only be identified as within his particular family and social network. These are the province of sociologists and anthropologists, so the book has kinship with them. Naturally also, as our book is about development, it has education as a close relative. A specialist in any of these disciplines will easily recognize the debt we owe him. I beg him too to bear with the elementary nature of what has been borrowed.

Life Situations and the Developmental Point of View

Returning again to the examples of our stone age man and modern young mother. The description of the man tells us little about his own personal life. But even the brief description of the young mother 'gets inside' her much more personally. We know she is married, her husband goes to work, she presumably does not earn money, she has at least three children, two of school age and one a baby. She continued her educational till eighteen or so. She has a lot of things to co-ordinate at once without, apparently, much outside help. She seems burdened by it all and can get angry.

9

We are beginning an account of one young mother's *life situation*. We see her tussling with desires and aspirations from within herself seeking to satisfy herself and those in the social network around her, in this case her family. Her mind dwells upon the near and more distant future, tussling with *conflicting* issues in a *dynamic* way. She is *problem solving*; it is part of any ordinary life.

We certainly must develop knowledge like this if we are to be personally helpful to our friends. It is the need for information of this sort that makes gossip, of a decent non–malign kind, so important anywhere.

Ordinarily we pick up our knowledge of friends' life situations in this informal gossipy way. However everyone in the helping professions often needs to be much more systematic, careful and searching in the analysis of their clients' or patients' life situations, this both externally in their environments and internally within their emotional lives. Our book is a first step towards developing this facility. It is laid out in age phases from birth to death. This is partly to give an acquaintance with some of the common experiences of each phase. These will note the emergence of new patterns of mental activity which continue on in modified form and affect later life.

The chapters describing and discussing the earlier phases of life will at the same time introduce concepts which are of importance no matter what age is being considered. This is the *developmental* point of view. It proposes that, however mature a person's thought may be, there is nevertheless primitive thought and feeling still active within it.

Things To Do When Reading this Book

The book describes and gives some theory about life situations from before birth to death. Since concepts introduced early in the book may be used later, it is as well to start at the beginning and go on from there. It may not be very satisfactory to dip in here and there.

Remember what is written down is only one person's point of view, it is fairly typical for a psychoanalyst but it is not a gospel, it is limited in its data and vision. Think of it as asking pertinent questions, let yourself argue with it, think for yourself.

It is a warm up to reading more detailed books. There are thousands of other points of view, millions of other facts and a myriad more puzzles which still remain. This is an overview book, it goes into very little detail on each topic. The references will help you start further study.

At the same time, if you have the opportunity, talk with your friends and teachers and argue with their observations and opinions. It is a very good idea, essential even, especially when considering the early years, to observe, play with, or talk to those at the age you are studying. You will find that people usually like telling about themselves (Roberts & Tamburri, 1981).

I, as a writer, am no special authority on any age phase, though I have had quite a lot of acquaintance with people in all of them. So for careful study you must attend to the ideas of specialists. Remember that, since the book is not about abnormality, so the references are not either. At the end of each chapter is a short list for further reading on the subject. At the end of the book is a list of more specialized references which are also indicated in the appropriate places in the chapters. Some of these are source books, not necessarily easy reading.

Also at the end of the book, under chapter headings, are references to films and videotapes which may be of use to classes. These are of particular value, especially when studying infancy and childhood, they often have a veracity and immediacy that no book can match.

Remember, particularly that this is *not* a book about pathology, abnormality or handicap. If you are a worker in the helping professions many of your clients or patients will be much more disturbed or disadvantaged than the people quoted here.

You will be bitterly disappointed if you read the book hoping for direct help in coping with the dire problems you may have to face everyday.

I hope there is, however, a more general value in the book. Nothing can be understood in isolation. We can only know anything by comparing it with other things. It is important to recognize this particularly in the helping professions where concepts about illness and health are often confused. Remember that sick processes can only properly be understood by reference to healthy ones. This book attempts to give a conceptual background of ordinariness which is useful to have in mind when facing unordinary people.

FURTHER READING

1 Beckett, B. S. (1981), *Illustrated Human Biology* (Oxford: OUP). For those who need to brush up their biology.
2 Carter, N. (1980), *Development, Growth and Ageing* (London: Croom Helm). An alternative general text.
3 Hunt, S. and Hilton, J. (1975), *Individual Development and Social Experience* (London: Allen & Unwin) Another alternative which emphasizes the experimental approach.
4 Kahn, J. and Wright, S. E. (1980), *Human Growth Development* (Oxford: Pergamon). This alternative emphasizes education.
5 Miller, J. (1983), *The Human Body* (London: Cape). Likewise, for those who need to learn some simple anatomy.

CHAPTER 2

Being Pregnant

The Early Organizers of a Child's Environment

Before we begin thinking about a pregnant woman directly it would be a good idea to dwell for a moment on the infant in his physical and social environment. He must from the beginning exchange substances with his environment. But his central control system, which gives him the capacity to discriminate and react selectively to the world of objects around him, is still only rudimentary. He is incapable of surviving alone. He must attune to organizers, his parents in particular, and they must attune to him to provide the substances and patterns of stimulation necessary for his survival. Only slowly is the necessity for them outgrown, even then early memory of them seems to be, in many aspects, indelible. It is one of the main contentions of this book that, just as genetic constitution delimits the pattern of a life, so early external organizers can delimit too.

In our culture and in many others the infant usually has one central organizer. This is most frequently his mother. Naturally this central organizer may be a foster, adoptive or grandmother, a nanny, wet nurse, or, more rarely, a father. There may be no one central organizer but several. Here we will be concerned with the preparations of a biological mother before birth.

The Nature of Preparation

Let us consider the nature of preparation generally for a moment, not just in pregnancy. To be conscious of preparing we must be aware, if only in dim or distorted form, of having a self with a body in a world of other selves and bodies. We must also anticipate the future in our imagination. This may depend on memories of similar happenings in the past but, since the future

13

cannot be certain, much anticipation must rest on inexact foresight. If we anticipate fulfilment in imagination, we usually feel *confidence*. But when we imagine or fantasize failure, we experience *anxiety*. It is common to avoid, or flee in imagination from, anxious anticipation. A pretence like this which gets away from feeling frightened, or distressed in any way is called *defensive* mental activity. These are 'cowardly' tricks but we are all prone to and need them. Much more will be said about defenses later in the book.

When we imagine the possibility of both fulfilment and failure yet persevere with a task we are being *courageous*. And if, when aiming for something, we *discipline* our imagination so that it represents *reality*, then we are problem solving. You will see, we are back to the psychodynamic view mentioned in the last chapter.

A woman's pregnancy is full of anticipations of problems to solve. Let us now begin to put ourselves in her shoes.

Conception

People who want to do something personally about continuing the human species, even in these technological days, usually have to start wth sexual intercourse.

Equally obviously, humans have found that biological reproduction is not the only fulfilment to be found in love making. Of all the hundreds and thousands of time a man or woman enjoys making love during his or her lifetime only very rarely do they want to conceive. Perhaps even more rarely will they want to conceive *and* rear their child. What is often unrecognzied is that some women deeply wish and will go to great lengths to be impregnated, while at the same time recoiling from the idea of mothering a child. Likewise, many men will go to similar lengths to impregnate without wanting to really father their child. The difference between the fantasy of procreation, stirred no doubt by physiological drives together with anxious competitiveness about fertility or potency, and the years of devotion entailed in rearing viable members of the next generation, is very great indeed. Let us look at the first months of these years of devotion.

Early Pregnancy: the Mother's Experience

In utero the human life proceeds from one cell to an infant of 2 million million cells in the course of nine months. This entails enormous chemical, particularly hormonal, changes in the mother's body (Chamberlain, 1969; Kitzinger, 1980). The manifest symptoms of these changes will be a woman's first indications of pregnancy. Briefly these are as follows. She will miss her period, will probably feel unusually tired, notice tingling in her breasts, probably suffer bouts of nausea, her vulva, nipples and cheeks will appear darker, she will need to urinate more frequently and she will put on weight. With a first pregnancy a woman will usually need confirmation from a doctor. When it is confirmed she very often finds herself swept by such feelings as, 'This is what life is all about'.

It is this profound feeling of fulfilment which, no doubt, makes an unwanted pregnancy and abortion such a disturbing business quite apart from its shame and guilt. The destruction of a new life is very frequently deeply mourned after abortions. This depression can happen to men as well as to women though less strongly.

Returning to a wanted pregnancy, one of a woman's first impulses is usually to share the news with others. She particularly wants affirmative enjoyment from her husband, family and friends, partly, no doubt, because they are going to be needed to take on responsibilities. Being pregnant without this gladness on the part of husband and relatives is nearly always a very lonely experience. There is little definitive research which correlates this loneliness with physical disorders for either mother or infant. There is only some evidence that stress generally can affect the foetus (Hunt and Hilton, 1975). But health visitors and welfare agencies, who have worked with isolated pregnant women, will usually affirm how much friendly help is needed. Being pregnant is a social event from the beginning.

The Emergence of Fantasy in Pregnancy

With widespread hormonal changes and excited yet anxious anticipations of an uncertain future, the old ways of a woman's personal adaptation tend slowly to be broken down. This change is subtle and does not become obvious until late in pregnancy, but it usually starts early. There is a proneness to lability of mood, to swing between elation and tearful depression with unexpected vivid thoughts suddenly coming to consciousness. A woman often has quick and idiosyncratic likes and revulsions about drinks, foods, smells or sights. Usually, if they are not too anxious, women need to be left alone to keep these ideas private. Her friends may chuckle at how peculiar her whims are at times.

There isn't any doubt that imagining or *fantasy* is going on in these moods. It is plain, too, that fantasy is often a close mental representation of *bodily functions* (Segal, 1985). Thus, for instance, we have just noted that pregnant women have impulsive thoughts about drink, food and smells. These are oral fantasies breaking through the usual ways of adaptation of the central control system, or *ego* as it is often referred to. The actual conscious images or intellectual elements in much fantasizing is often minimal or non–existent. When they do emerge they are normally confined to visual dream–like symbols that are thinly disguised portrayals of physiological events. Here are some examples given by pregnant women.

I suddenly had the idea of a concrete mixer being fed with ice cream.

I had the thought of a waterfall in a wooded cleft in the rocks.

A woman took her parents out for the evening. She felt she was very insensitive to their needs and reported that she felt 'like a lump–headed idiot'. The following night she dreamt she 'had a tumour on the brain'.

In this last example the woman has a psychological experience with her parents which she expresses by using a metaphor. It is turned, more malignantly, into a physiological event in her dream. Digressing a moment to a bit of general theory. Although this notion of fantasy is really a simple one, it is often particularly difficult to grasp because we are so used to thinking about tasks rather than moods and bodily rooted feelings. Perhaps a better term than fantasy would be 'feelthought' to indicate the influence of body feeling and ideation implied in it. This concept of fantasy is not on the whole subscribed to by some academic

16

and research psychologists. Behaviourist psychologists for instance tend to reject it because its occurrence is not verifiable by observable behaviour. I think that they are the poorer for this because fantasy seems to be an omnipresent aspect of experience. Others, like the child psychologist Piaget, are interested in fantasy but tend to ignore the personal feelings involved. Psychoanalysts and therapists consider them to be of prime importance.

Returning to our pregnant woman, the direction of her fantasies and also her adaptive thoughts preparing for the future, will depend largely upon the nature of the society around her. Let us take some of the facets of this in turn.

A Woman and her Culture

The pregnant woman's feelings and anxious fantasy will naturally depend upon the general expectations of her culture. For instance in many cultures, especially in the past, it was generally felt good to be fruitful, hence easy to be joyful about pregnancy. But many societies valued male children far above female ones. In such climates it is to be expected that women would be more prone to worry about the sex of the infant than they do in, say, Britain today. In this country, where birth control usually is felt to be a necessity, a woman is more likely to be worried about whether she ought to be having a child at all.

A Woman and her own Mother

One common trend which runs through many women's doubts about themselves is implicit comparison and rivalry with their own mothers. Such feelings are not often consciously expressed, but when they are they have a poignancy which rings true. For instance:

> My mother believed that childhood was the golden age and a mother must surrender everything to the children. I don't want to do that, but I am terrified not to.

> Mother never lost her temper with us. I am sure I shall. I am awful, aren't I?

17

Mother always made us go to the lavatory every few hours; I am determined that my children shall not suffer such indignity.

It seems that pregnancy usually involves a complex of feelings of conflict, rivalry and conformity between the generations (Clulow, 1983; Deutsch, 1944; Oakley, 1979). It is nearly always very important to a woman that her own mother should be pleased and appreciative of the pregnancy. For instance:

A young woman who was a very active and convinced rebel against the absurdities of the older generation said some months after her baby was born. 'My mother is a crazy idiot, but I forgive her everything and love her dearly because she dotes on Ann, the baby. I could murder my mother–in–law because she has just blatantly ignored us since I got pregnant.'

It is possible to detect a thread of anticipatory thinking in most of these comments about mothers. The women seem to be forming resolves about how they will behave as mothers and comparing these with their own, perhaps distorted memories of childhood. This, incidentally, can be a most enjoyable activity. It also seems that such anticipatory thinking serves definite preparatory functions. It helps a woman to form a preliminary attitude towards handling a baby. This in itself is valuable because it helps her to be certain of what she feels to begin with. Then, depending on how things turn out later, she can alter her ideas to match her experience without too much floundering. If she does not have any preliminary attitude she is much more likely to feel lost and helpless. Here is a sort of hypothesis testing or informal scientific method.

A Woman and the Professional Helper

Most women want to get on with being pregnant with the minimum of interference from professionals. At the same time they anxiously rely on their physical examinations and advice, and often feel very sensitive to being in their power. Some women find it easy to be in such a state of helplessness, others are terrified and enraged that they are treated like cattle. This is not so often experienced with a home confinement, which, however, is increasingly rare these days. Long waits and peremptory treatment by busy staff in hospitals has broken the calm peacefulness of many pregnant women. On a more personal level, a few midwives have been felt as the terror of the neighbourhood, leaving young mothers weeping helplessly in fright, hating them yet not daring to say anything because they will be in their hands at the birth. This does not diminish the ordinary midwife, rather it shows how very important she is.

We have no way of estimating the lasting effects of these bad experiences. They are possibly not very great, but the more time taken up by being anxious and angry means less just relaxing and being with her baby inside her. If she has other personal anxieties as well, an unpleasant professional helper may be the straw which breaks the camel's back. On the other hand a good, friendly midwife or Health Visitor is an enormous help to a worried young mother.

A Woman and her Husband

Although some women today intentionally set out to be pregnant alone, the person who is still usually most important for a woman's ease in pregnancy is her husband. With him there is the obvious, yet fundamental, need for assurance of financial and physical security and for intimate valuing of each other. Lost intimacy can act like a malignant thing.

A woman will often say how sensitive she is to her husband's moods. It matters very much that he should want their child, for without this a woman is alone and often feels guilty both towards her husband and the baby. Even though he wants a child, it is not uncommon for a husband to be physically disgusted by his pregnant wife, and when this happens she is likely to become listless and depressed. When the baby is born, the husband's disgust may wear off and the wife's depression lift. Probably nearly all women worry that they will be sexually unattractive to their husbands during and after pregnancy. Love-making will inevitably be interrupted at the end of pregnancy and immediately after birth. It may even be many months before easy intercourse is re-established. Husbands often find this abstinence difficult to bear and become impatient and bad-tempered, perhaps because they have not anticipated these difficulties. It is quite remarkable how many men choose this time to have an affair or to be cruel in one way or another.

The importance of the feeling of togetherness is reported so often by women that it is clearly one factor which can shatter the ease of a woman's maternal preoccupation. It is often quite sufficient for a woman to feel that her husband is with her in spirit. She can be quite happy if her husband is thousands of miles away if assured of his regard. Nevertheless pregnant women do repeatedly stress the value of day–to–day contact with their husbands. This has probably become more important in recent years as husbands and wives often have to support each other rather than being able to turn to other members of an extended family.

One responsibility of a husband in day–to–day contact is to act as a receptor and container for his wife's anxiety–ridden fantasies. How one person helps ease the anxiety of another in this way is not yet well understood. One instance of it would be when a wife worries that the baby will be deformed and her husband sympathizes with her, but points out that it is unlikely. Naturally enough, husbands can also increase anxiety both by panicking themselves or perhaps by being disdainful or finding one of a thousand ways of being nasty. In whatever way a husband responds, his wife is likely to be sensitive and easily stirred by him (Clulow, 1982; Oakley, 1979).

The Husband's Experience

At this point, put yourselves in the husband's shoes for a moment. It does not take much exploration to discover that his wife's pregnancy usually brings to a husband a sense of fulfilment and passionate pride. This is very close to the emotions of his wife so they can intimately share in each other's feelings.

Just as for his wife, with her pregnancy he is called upon to develop, to change his old ways of life and thought, though to a less obvious degree. Here are a couple of examples.

After that first feeling of 'Ah, we've done it', I began to feel scared. Would my wife be all right? What would I be like as a father? And then all sorts of worries about whether our housing would be all right, would I earn enough money to keep the family, came upon me.

Money seemed my main worry. My wife had worked before and now it was up to me – quite exciting in its way but scaring all the same.

A very common anxiety for a man is that his wife will withdraw from him and not dote on him as before. Probably every man wants to be mothered, and looks to his wife to meet this need. With a real baby as a rival this satisfaction is threatened. Here is an extreme example.

I know I feel an outsider. My wife says I make myself one. But I know something disappeared for me when the first child was born which has never returned. And the second child seemed to finish things. I am very fond of both children but something went dead between me and my wife.

You will notice a sense of puzzlement in this report. Something is happening *inside* this man which he does not understand; it seems that an *unconscious* mental process is occurring, we will refer to this sort of thing many times in the book.

Returning to a husband's alienation, perhaps this is a common root in the cruelty which we noted as being remarkably common a few paragraphs back. But for most men alienation is probably hardly felt at all. They shift happily to sharing. But it seems rare for a man never to feel afraid of being left out, if not in the first pregnancy then in later ones. In some ways it is inevitable, because a woman turns into herself and it would be an insensitive man who did not feel it. Thus in pregnancy when the wife needs her husband's concern perhaps more than on any other occasion, this is nevertheless the time when he is very likely to feel alienated.

As with their wives, pregnancy stirs many men to fantasies. These often have their roots in childhood just as those of their wives do.

My parents counted up our misdeeds until the end of the week, when we were beaten for them. I am determined never to lay hands on my children.

I find it easy to look after children. I suppose I got it from my mother, who was always a nice, warm person.

My father enjoyed playing with us and I am looking forward to doing the same.

21

And, just as with the wife, such ideas must usually remain private to himself and are not often discussed, except perhaps with her (Cath, 1982; Parke, 1981; Rapoport, 1977).

Later Pregnancy: Changing Image of the Self

At about four or five months a woman will first notice the *quickening*. This is the flutter of a baby's *movements in utero*. It is usually very exciting, a living thing which is not herself is inside her. She frequently starts up imaginary conversations with her baby, often giving it provisional names. The actual anticipation of birth and mothering is likely to become more and more omnipresent in her mind. At the same time she is getting heavier and thus less able to move with her old vivacity. If she is going to relax and enjoy the pregnancy and early mothering a woman just has to give up many of her old ways of organizing herself and her life. The process of allowing this to happen has been termed *primary maternal preoccupation* (Winnicott, 1958). This summarizes many of the changes of pregnancy mentioned already: becoming slower in movement, a pleasant withdrawal into ideas of self and infant, opening up of fantasies, loosening of old habits, and a *regression* into allowing the self to be dependent in more childlike ways than previously. The easy functioning of this process is epitomized by quietness.

The fact that many adoptive mothers both enjoy being and make themselves enjoyable mothers makes it plain that the physiological basis of this primary maternal preoccupation is not necessary for good mothering. But I think most experts on early motherhood would agree that it is really very important, especially when 'heavy with child', to let old timebound habits and vigilances relax so that maternal preoccupation can take its course enjoyably.

Since preoccupation involves the breaking of old modes of adaptation allowing new modes of activity to be tried and tested, later pregnancy is a major developmental transition or *life crisis* (Caplan, 1964). This concept of crisis will be discussed on several occasions later, particularly in Chapter 9 on 'Adolescence'. Pregnancy is not usually a noisy crisis and is most frequently benign but, as it is a time of fundamental life change, it is worth regarding it as a crisis so long as we recognize that it is normally happy. And as with any crisis anxiety is more or less inevitable. Furthermore, in later pregnancy, as a woman cannot both relax into preoccupation and remain fully vigilant, she does need to depend upon others to be vigilant for her. If her husband cannot do this then she needs other helpers.

Let us now consider a few anxious ideas at this time of preoccupation.

A pregnant woman often feels that she will not be able to cope with her new life. There also comes rage at having to change her old ways. A common occurrence today is that a woman may have invested a great deal in a career, identified with it, and gained a unique sense of self–esteem from it. The coming of a child threatens this, and it is often possible to see how a woman cannot let herself relax into maternal preoccupation because she needs to cling to her career and old sense of herself. This affects some women so deeply that they seem to deny their pregnancy for as long as they can. Then, when they can deny it no longer, they become angry, tense and depressed.

A piece of advice about going back to work is often given but, I think, far too little heeded. It is, 'If you possibly can, don't make any commitment about going back before you have had your baby. When you're pregnant it is not easy to judge, at least wait till you've tried it with the baby.'

If it is necessary to go back to work soon after the birth then, as we will see in the next chapter, it is essential to be freed from all outside timetables during the first few weeks. Mother and baby must have time to attune to each other. Much needs to be done in our society to give more flexibility to mothers, but careers and jobs must wait in the first instance if a woman is going to enjoy her baby.

These heavy judgements are not made from the point of view of a male chauvinist who says a woman's place is always in the home. In Western culture, where individual freedom is well valued, a woman's work and career is deeply important. But if a woman has chosen to have a baby, and it is nearly always a choice these days, she is asking for trouble for herself or her baby, if she thinks she can go on with her career quite uninterrupted (Leach, 1974, 1983; Oakley, 1979).

Less tied to external circumstances, and hence more difficult for outsiders to comprehend, is the problem of lack of self–esteem which all mothers experience to some degree and some feel chronically. Some women seem, as an underlying characteristic, to doubt themselves as having any value, so that when they become pregnant they have not got the self–confidence to say to themselves, 'Now I can relax, I am good enough to be a mother and I shall know intuitively what is right'. For instance:

A woman was convinced her own conception had been a mistake. Her parents had spent all their time running a shop, so she lodged with her grandparents or aunts throughout most of her childhood. She was swept by anxieties during all her pregnancies, obsessed by the thought that it was all a mistake, that she didn't want the baby and couldn't cope. However she repeatedly got herself pregnant, perhaps to prove that the next one would be wanted after all.

Such an illustration cannot be a convincing general proof of the importance of early childhood experiences, but it alerts us to their possible importance. This is given added support by the observation that most women who felt predominantly at ease with their own parents as children tend to feel serene about their own motherliness as long as they are happy with their husbands. I know of no formal survey that confirms or denies this, but it is a common enough experience of those working with pregnant women.

Perhaps the most often felt anxiety about mothering is concerned with aggressiveness in one form or another. A mother usually expresses this as a fear of losing her temper, or of not being able to stand up to her child's demands. Many women feel they should show no aggressiveness towards their children and suffer agonies when, out of frustration, they quite naturally find it rising up in themselves. Remember here that a show of aggressiveness helps a child discover limits for his behaviour. Fear of aggressiveness seems to vary from person to person, from family to family, and from one culture to another.

Probably in Britain, where social norms are more fluid than in traditional societies, a woman is more or less allowed the freedom to find her own way with her children. But this freedom of choice brings its own troubles, for it means that a great burden of responsibility and hence guilt rests upon her shoulders. If anything, the burden has increased with greater understanding of the needs of children. To many mothers the professional adviser has something of the aura of the priest of old who purveys gloomy knowledge of what is right and wrong. Responsible parents cannot ignore technical understanding of the needs of children. But if they blindly obey the conflicting pronouncements of every expert on the subject they are likely to be driven crazy with doubt. One must just be ready to take one's own decisions.

The Days Before the Labour

In the last weeks of pregnancy there are a lot of household arrangements that need finalizing, getting a place for the baby, his cot, clothes, nappies, bath, pram and so on. It is usually great fun. There is also arranging with those who might be coming to help after the birth. All this means working out living arrangements in advance, there is no time to do it after the baby is born.

At the same time there will probably be a lot of anxious excitement about the labour itself. Preparation for child-birth in classes is now a well established part of health care. A great deal of thought and research has gone into this. One trend resulting from this has been towards greater technical intervention as medical knowledge has increased. Thus for instance in Britain today virtually all first births are planned to take place in hospital where emergency equipment is readily available. The other trend is towards women becoming more active participants in their childbirth. These two trends often conflict. We cannot enter this debate here, but any one directly concerned with childbirth cannot avoid it (Kitzinger, 1980; MacFarlane, 1977).

Second and Further Pregnancies

Much of what has been said about the first pregnancy applies equally to later ones. There may not be the extremes of pleasure and fear which can be experienced with any event for the first time. Both parents will probably be surer of themselves, at least if all has gone well with the first child. In a sense the first child, simply by being alive and well, contributes to the parents'

25

confidence. This will not be by any conscious intention of unselfishness or kindness on the child's part. But it is communicated in a very real sense none the less. Winnicott (1964) has called this a child's *contribution*.

Just as parents have to change when they have a baby, so also the first child will have to accept the change when the second one is on the way. Most children are very aware of the changes that take place in their mothers in pregnancy. This will be perceived in her body shape and also in her tiredness and withdrawal. It is usually possible to see signs of anxiety about this if one listens and watches the first child. It is not something a child can easily talk about and being unable to communicate makes it all the worse for him. The child tends to be in the same position as the husband whom we described above, alienated from his mother, who, because of her pregnancy, is not able to respond to him as sensitively as before. It is noticeable that many kind and devoted parents ignore this pregnancy anxiety of the child. It is often thought that the child will be jealous of the baby after the birth. This is very likely to be so, but anxiety before birth, related to the mother's withdrawal, is also most common. It is often thought that because a child is too young to understand the facts of life then he will not not be frightened about the changes in his mother. However the younger a child is, the less he understands, and hence the more puzzled and frightened he can be by the mysterious differences in his mother. Research evidence shows that even very young infants are highly sensitive to their mother's state. Leaving this unrecognized spells trouble for the future (Bower, 1977, 1982; Lewin, 1975; Leach, 1974).

Summary

This brief discussion of a woman's and her husband's feelings and fantasies gives us few definitive conclusions except that of the importance of 'vigilant helpers' who will allow a mother to be preoccupied in a relaxed way. It also demonstrates how pregnancy, like the rest of childhood, is a private matter of each individual's fantasy and problem solving. But it is also a public process where the private activities affect all other individuals involved in the business of bringing up a child. This web of related private worlds makes the *psychological environment* for a new born child.

FURTHER READING

1 Chamberlain, G. (1969), *The Safety of the Unborn Child* (Harmondsworth: Penguin).
2 Kitzinger, S. (1980), *Pregnancy and Childbirth* (London: Michael Joseph).
3 Smith, A. (1970), *The Body* (Harmondsworth: Penguin).

CHAPTER 3

The First Months

Observing Young Babies

We now need to start feeling along with little babies as well as their parents. This identification is not very easy because, of course, babies can't talk. We can however, deduce a great deal from careful systematic observations, greatly assisted in recent years by videotaping and other forms of recording. Remember it is important to observe babies for yourselves. Even if you can do no more, a few minutes comparing the facial expression of a baby of a few weeks with those at a few months will make you marvel at what happens in those early weeks of life. It is very useful indeed to fix up with a young mother to observe a baby systematically. Failing that, as was mentioned earlier, there are very good films on the subject (Sunday's Child for instance: see film index), or follow Roberts' and Tamburri's (1981) observations. The least you can do for yourselves is spend a bit of time looking at babies in their prams around the shops. As long as they are reassured about baby snatching, mothers usually love their babies being of interest.

Labour and After

The first stage of labour, as it is called, involves the opening of the cervix between the uterus and vagina. This is effected by the progressive contractions of the cervical muscles, labour 'pains' derive from fatigue of these muscles. For some mothers this stage is quite brief, an hour or so. For others it can last for many hours; being very painful and exhausting. Many influences no doubt contribute to the ease or otherwise of this stage (Breen, 1975). The childbirth classes referred to in the last chapter are directed mainly to facilitate this first stage (Kitzinger, 1984; MacFarlane, 1977; Oakley, 1975; Richards, 1980; Yarrow, 1977).

The Second Stage is the actual passing out of the baby from his mother. It is wonderful to see; quite quickly, after the hours of the first stage, his head will appear and, once the head emerges, the rest usually slips out easily enough. Within a few moments, stimulated by the lack of oxygen in his blood stream, because it is now not coming through the umbilical cord from his mother, the baby coughs, splutters and begins to cry. He is breathing and turns miraculously from a blueish to a full pink colour.

For a baby, birth means a fundamental change in his whole psychological organization. *In utero* ventilation, nurition and excretion took place through the placenta and umbilical cord. Now ventilation starts suddenly with the first breath through the lungs, feeding by sucking commences within a couple of days and so also does bowel and bladder excretion. None of the reflex patterns organizing these functions can have been fully exercised before.

It is common practice now to give the baby quite quickly to his mother to have a cuddle together. She needs to get the feel of him outside her after months of growing inside. For him it will be his first sight and smell of anything. What this is like for him is hard to imagine. We will think about it more later (Klaus, 1976; Leach, 1974, 1983; Slukin, 1983).

The early hours of life are usually spent asleep, preferably wrapped up tight, as near as possible to the all round support of the womb. A mother is most likely to be exhausted and glad of the baby's sleep in order to sleep herself.

Within a few hours the first steps in establishing a feeding routine will have begun (Kitzinger, 1979). It is a day or so before breast milk begins to flow. But a baby is usually put to the breast before this to exercise his sucking reflexes and to stimulate production of the mother's hormones which themselves start milk production.

Even if feeding is by bottle from the start, the attuning of the baby and mother to each other is very important as we shall see later. The patterning of innate reflexes, is highly complex even for simple feeding and they need exercise to operate efficiently. A baby will probably have learnt something of his mother's movements and rhythms, like the sound of her voice while *in utero* (Verney, 1982). But the early days and weeks after birth have to be taken up with the two of them *settling down* together.

29

To begin with, although you will see that a baby has movements and physical characteristics that are very much his own, his activity, rhythms of feeding, excreting, sleeping and wakefulness are very unpredictable. For his mother this randomness is very exhausting, being up any hour of the day or night, trying to satisfy him. The very weariness and anxiety about whether all will be well is depressing. The local Health Visitor can be an invaluable adviser at this time.

Our mother is having to get over her own pregnancy and labour and at the same time attune herself to her baby's rhythm, which often is so random. It is this need to attune that makes maternal preoccupation so important. It becomes plain why a mother should have no fixed outside time table. Her mothering will have to be taken over, more or less, by someone else if she insists on keeping up many outside activities.

In time, for most sooner rather than later, a rhythm of feeding and sleeping will establish itself, the baby begins to get hungry and wakeful at more or less predictable times. Naturally this will very largely be due to flexibility on the mother's part, but some physical *negotiation* must go on. For instance she will discover when to get him to wait a bit without too much trouble and so on. After pregnancy and labour, establishing her baby's rhythm is a woman's first triumph of motherhood (Clulow, 1982; Oakley, 1979; Schaffer, 1977; Stern, 1977; Winnicott, 1964).

Sleep Arousal and Distress

It is useful here to digress to introduce some ideas that have relevance beyond infancy. As we have just seen, a baby moves in a cycle between sleep and arousal; if disturbed in this he becomes distressed, this is usually signified by crying. These three stages, sleep, arousal and distress, involve definable differences in the activity of the nervous system of adults as well as infants. These are important to recognize when considering any sort of stress, anxiety, anger, defence and aggression, as we shall be doing throughout the book. Early recognition of this pattern was made by the physiologist Cannon and has been followed by much research since.

A summary of each state in turn is as follows.

(1) *Sleep*. There is: synchronous activity in the central nervous system (CNS); inhibition of the sympathetic nervous system; and activity of the parasympathetic. Of subjective emotion there is none (except when sleep is disturbed).

(2) *Arousal*. There is: partial desynchrony of the CNS; a patterned mixture of inhibition and activity of the sympathetic and parasympathetic systems. And subjectively we experience interest ranging from pleasure to anxiety.

(3) *Distress*. There is: desynchronous activity of the CNS; sympathetic nervous activity and parasympathetic inhibition. Subjectively the experience is extreme rage, fear or excitement with hyperactive behaviour and hypersensitivity.

Distress is thus characterized by nervous disorganization and hence mental disturbance. Also present are elements, often in confused form, of animal *flight* and *fight* reactions. These in turn affect breathing, temperature control, digestion and excretion. Distress is a *psychosomatic reaction* (Parkes, 1972).

What evokes distress in young infants? Well known for many years have been the following: sudden loud noises, loss of body support, choking, hunger, pain and cold. Less well known and recently added are: *over–stimulation* (e.g. erratic jigging which disturbs the sleep–waking rhythm), *confusion* (e.g. after the baby is a month or so old a strange voice interposed with his mother's face will distress him), and *rejective gestures* by a familiar person like his mother (again this occurs only after there has been enough time for a person to become familiar). Distress is relieved, of course, by stopping the disturbing stimulation and also by nursing, holding, caressing and cuddling. These all involve *gentle* movement and skin sense stimulation. They are *gently erotic*, rather than excitingly so, and have more than superficial similarities with a lover's embrace. That young babies actively seek this comfortable pleasure, not only when distressed, is obvious enough to any interested mother. The necessity for this has been systematically studied in primates. But until recently the importance of erotic holding and play has been largely ignored by academic research, which has concentrated on more intellectual development. Psychoanalysts, on the other hand, while backward in investigating intellectal discriminations, have long stressed the importance of erotic comfort (Bowlby, 1969; Winnicott, 1965; Escalona, 1969).

31

Certainly sleep for both a baby and his parents is vital. It is inevitably broken by night feeds, but a little baby who can't go off to sleep is dreadfully pathetic and seems catastrophic to his mother. The trouble can often be resolved by working out some emotion laden habit that his mother has got stuck into and which keeps him aroused. A few visits to an expert on sleep disturbances at a paediatric department or child guidance clinic can work wonders (Asher, 1984; Douglas and Richman, 1984).

What Can a Baby Do in the Early Weeks?
(Bower, 1977; Shaffer and Dunn, 1980; Sylva and Hunt, 1979).

It used to be assumed that a baby did very little except use rudimentary reflexes in sleep and when feeding in the early weeks. There is no doubt that sleep fills very much of his time. Development of nerve tissue as well as other organs is incomplete at birth and even for months after. How much this development is dependent upon sleep is not clear.

Recent recording methods have made it clear that wakefulness and articulated awareness is much more manifest in earlier infancy than used to be thought. For instance, it has been found that, until about two months, a baby is very shortsighted with a focus at about 20 cm. (8 in.). As most objects are outside this range they must be a blurr to him, and he will appear vague or uninterested to the more distant onlooker, but he is fully awake. If one brings objects close up he very quickly gets intrigued, particularly by the textures of things. He will begin excited reaching gestures as early as two weeks old.

At a few weeks he will stop crying at the sight of his mother's face, but not at others. At six weeks or thereabouts he will smile for the first time (Bower, 1977, 1982; Leach, 1974, 1983; Spitz, 1965).

One of the beginnings of the sense of *distance* is manifest within a month. Thus a 'defensive' gesture is made with the head and face when an object comes within about 30 cm. (12 in.). Objects at a greater distance do not evoke this gesture even though they are bigger and make a larger retinal image. Odour discrimination is acute very early. Mouth, nose and eyes are the focus of most of the baby's organized activity with his environment. The sucking reflexes become more assured and also more adaptable over the days. It is no wonder that psychoanalysts have called this time of life the early *Oral Phase*.

We noted reaching gestures with the hands as early as fourteen days. Hand–mouth–eye co–ordination grows apace so that by three months old hand movements and touch are clearly, even easily, being articulated to what is visually seen. Even before this we can see the beginnings of *manipulative play*, twisting turning and moving things with the hands which are also looked at, taken into the mouth, chewed at and smelled.

Without things to look at and play with a baby of a few months old displays every indication of *boredom*. He needs quiet unstartling things to look at, handle and get to know. There seems to be crucial age phases for the initiation of each new co–ordination. The evidence seems to be that without gentle stimulation children miss out developmentally (Leach, 1974), 1983).

Although we now realize that a baby is much more intelligent than we used to think, nevertheless the world around must be very puzzling for him. For instance, although grasping is well co–ordinated by about six months, a baby will grasp at such intangibles as soap bubbles, sunbeams and also bright pictures at distances beyond his reach (Bower, 1977, 1982).

Adaptation, Sameness and Similarity

With this example of grasping at soap bubbles in mind let us turn for a few months to think theoretically about an infant's adaptive mental activity. We cannot call it proper thinking yet because he is not actively and systematically using mental representations of the world in his mind. But the preliminary stages of *adaptation* are appearing. We have already noted that the term used generally to refer to an individual's processes that adapt to the world around him is *ego–functioning*. It can apply to anyone of any age. Psychoanalysts introduced the term ego to refer to those aspects of the mind that are adaptive in contrast to a person's drives and desires which are termed *id*. (Freud, 1923). The ego should also be distinguished from the *self*. The two ideas are related but the self refers to one as an *experiencing* subject whereas the ego refers to one as adaptive.

The infant, with his rudimentary ego–functioning, is just beginning to know what to do in the world. He is starting to register and memorize relationshps between things that are detectable within and around him. He seems to store them in organized systems, or in *schemata* as they are called (Piaget, 1953). For instance, with breast feeding say, the sight of a certain face (his mother's) is followed for him by a pink thing with a reddish target shaped blob in the middle that also has a certain smell and softeness. When sucked, liquid comes from the red thing and makes him feel better. Soon a baby, when hungry, begins to root with fingers twitching and sucking actions immediately he sees that face of his mother again. The sight of the face has begun *to remind* our baby of, or to *mean*, feeding. In this case one thing (face) is just beginning to be a *sign* for other things (breast, feeding).

Here our baby, using memory, is beginning to make a *same again* response. 'That's the same face, it means the same blob which means that same blissful liquid stuff again.' Likewise with other things, at bath times he learns to splash. In words this would be, 'Ah same again (water) it's splashable'. Or a familiar toy is seen as the 'same again' as grabbable and holdable.

The use of *memory* in the registering of *sameness* is, of course, one of the most widespread and fundamentally adaptive functions of the mind. We have here the rudiments of *classificatory activity* which is used for economy of thought and action, from the simplest to the most complex and abstract.

Of course infants and young children largely have to learn what things are the same as others and they often get it wrong. For instance, our baby grabbing at soap bubbles sees them in front of his eyes and thinks, 'Ah the same grabbable things again', when they are not.

Note also that registration of sameness probably must *precede* the recognition of *similarity*. To register things as similar involves knowing that they are *different* but have an attribute in common. Only after this does classification proper occur. Our baby has nowhere nearly reached this complexity of thinking yet. In fact Piaget noted that even his eighteen month child, out for a walk, saw a slug and then another one some minutes later. From her conversation it was clear that she thought they were the *same* slug. The idea of similarity had not occurred to her yet, at least with regard to slugs.

As we go through the stages of development we will see how the element of sameness, that seems to exist in any recognition of similarity, can emerge in its primitive form at any time of life. We will see that this happens most obviously in madness, but also quite normally when strong emotions are loosened (Matte Blanco, 1975).

Sight, Touch and Sound: Synchrony of Baby and Mother

Young babies are soon capable of a rough localization of sound and respond differently to its variations. His startle reflex to loud noises has been known for a long time. Only recently discovered, however, is that the whole body, and mouth in particular, responds in minute, but quite specific ways, to different patterns of sound. There is evidence that he *mimics* his mother's spoken mouth movements with similar 'linguistic' mouth movements of his own by the time he is a month or so old. This is matched by mimicry of other body movements by a similar age. For instance, he is able to respond to his mother's smile with the beginnings of a smile of his own. These observations make it clear that the rudiments of echoing another person's pattern of muscular activity, by movements of similar muscles, is present very early and probably has an innate basis. The clarity of this mimicry seems to get confused later in infancy, presumably through a baby's proneness to adapt to many external stimuli at once. What is more the mimicry is only brought out in the first place by an adult who *fits into* his arousal pattern.

The person most likely to fit like this is, of course, a baby's mother. It is clear that mothers and babies develop a *synchrony* with each other from the earliest weeks. As we have also seen, within a similar period the baby can discriminate his mother's from other faces. By this discrimination and synchrony *conversations* between mother and child establish themselves within a very few weeks. These conversations often involve sound, not meaningful words, of course, but rather visual–muscular and auditory *gestures*. Such gestures must, together with feeding experiences, form the *roots of human communication* and hence of language (Bower, 1977, 1982; Schaffer, 1977).

Baby and Mother Being Sociable Together

The idiosyncratic style of each baby and mother has already been stressed. Some babies are born very active, others placid; likewise mothers differ in their characteristic movements, not only from one mother to another but also from culture to culture. If a mother cannot attune to her baby's rhythm then as a result the baby will become distressed. This in turn stresses his mother, usually upsetting him more, so that a vicious circle is likely. Although babies are usually more robust than inexperienced mothers fear and 'forgiveness' comes quickly when a contented rhythm is re–established, nevertheless a mother needs all her vitality, ingenuity and patience to melt such vicious circles. It is naturally most enjoyable for both if these hardly occur. We have already considered that for this to happen a mother needs to become loose and attuned in a relaxed sort of way to her baby's rhythm and pattern.

This is why relaxing and regressing into maternal preoccupation is so important. A mother with only half her mind on her baby, or over quick and incisive in her movements, is unlikely to provide that smooth synchrony and moderate stimulation when he is awake and minimal stimulation when asleep which seem necessary for an infant's growth.

However it must not be forgotten that the mother–child relationship is a two–way process. From birth some babies seem to be impossible to please even with the most devoted mothering. And, even if this is not the case, a mother can only rarely be blamed for conscious wilfulness. When things go wrong a mother is, as often as not, as baffled and distressed as the baby.

Nevertheless, it is under these loose conditions that the mother–child conversations of feeding, cuddling, talking, smiling, and playing generally seem to come most easily. In them a mother seems to experience her baby paradoxically as separate from her and yet at one with her. 'Separate and yet as one' is naturally beautiful to watch. It is most obvious in mothering but also in fathering and in later life between lovers and very good friends; its beauty is also evident in anyone playing or working with consummate skill. This vital state of being together and yet separate is, I think, what we mean by *loving*. It is wonderful to experience and also seems vital to growth. However, loving can lose its balance: too much togetherness can become engulfment and separateness can slip into alienation. This theme of loving is central to the book and will be referred to time and again in different guises.

Returning to our baby, it must be becoming clear that, as he is awake little and cannot move himself, only very few people, one or two at most, can have the time to attune really harmoniously with him. What is more, if a mother herself is to relax with him, it is then necessary for her helpers to continue to be vigilant for her. This is where the *baby–mother–father triad* can be a most enjoyable and economical unit. A husband who is lovingly attuned to both his wife and baby can move easily between looking after her and helping her out with attending to their child. A husband is the most likely person to fulfil the needs of a vigilant helper, because not only are husband and wife usually close but so also are father and baby.

There seems to be no other reason why the nuclear family should be sacred at this time of life. Children thrive, in infancy at least, throughout the world with indifferent or absent fathers so long as other trusted helpers are available. Nevertheless, at this phase of a child's life, the reasons given so far alone seem sufficient to value and preserve the nuclear family as most precious.

The necessity for an infant to have only one 'mother', feeding her child herself with attendant father, has often been questioned. Successful adoption makes it plain that breast feeding is not essential. And the participation of nannies, or other members of an extended family, is common throughout the world without apparently catastrophic effects (Clarke, 1975).

Let us summarize the observations so far and come to provisional conclusions. They apply to babies in any culture.

Breast feeding in early infancy is optimal (but not necessary) in that it is a most mobile food factory, provides a balanced diet, and gives immunity to many infantile infections. It also provides the possibility of enjoyable erotic, visual and vocal conversations which do seem to be necessary for optimal development. These conversations may also be deeply enjoyable to mothers. But it is also plain that an infant can enjoy relating to more than one person so long as they are familiar and also attune themselves to the baby in a quiet relaxed way (Rutter, 1972). A chaotic whirl of unfamiliar helpers is known to be extremely distressing (Bowlby, 1953, 1969; Spitz, 1965). The grave long–term effects of this will be discussed in the next chapter.

In conclusion, it seems that a very small number of affectionate and familiar parents (biological or otherwise) are necessary for an infant's vital development. A biological mother breast feeding with close affectionate helpers is optimal. This is usually most easily attained, in Western society at least, in a nuclear family. These conclusions may seem severe, particularly on mothers who have to work. But with these thoughts in mind she may find it easier to come to decisions as to how best to balance work with looking after her child. Also, there is no doubt that much needs to be done to make work more flexible for men as well as women with young children. That said, if you have a choice why have a baby at all if you don't want to give him priority?

Forms of Pleasure

Using infancy as a starting point let us digress again to think more generally for a moment. We have already thought about arousal and distress. Now, within wakefulness we can distinguish several sorts of pleasure that can be enjoyed.

(1)　There is *homeostatic pleasure*. This is experienced when a chemical balance is restored in the body. Thus a thirsty baby is active, even distressed, until he has filled his stomach with sweet liquid. Then he flops back with a satiated sigh. A similar satisfied sigh is also noticeable after urinating or defecating. There seems to be something *orgasmic* about this satisfaction, in that irradiation of pleasure seems to shiver through the whole body (Winnicott, 1958). Perhaps the satisfaction of any biological drive is signalled by an orgasmic experience however slight.

(2)　*Erotic pleasure* is also distinguishable. Rhythmic stimulation of the skin and movement senses, such as being rocked, hugged, tickled, stroked and bounced, evoke paroxysms of laughing delight which have something of the quality of little orgasms. This pleasure is close to, yet quite different from the pleasure of *being held* which involves contact with little or no movement. In parallel with these, an infant also enjoys such self–stimulation as thumb–sucking. This is *auto–erotic* activity of which more will be said later.

(3) *Mastery and meaning–finding pleasure* is also detectable. Here there is delight when a new skill has been achieved, or some meaningful, new mental realization has been made. Most adults are familiar with the culmination experience, such as when a goal is scored or the shout of 'Eureka' when making a new discovery. Such pleasures, often accompanied by a laugh of triumph, again with a slightly orgasmic note of a mental kind, are easily noticeable in older babies.

(4) *Social or conversation pleasure* – this obviously often includes both erotic and meaning–finding elements. It is most vividly and easily seen in an infant in synchrony with his mother but, with good fortune we enrich ourselves all our lives with these human interactions.

Psychoanalysts have stressed the first two pleasures together with the fourth. They have developed theories centering about the fundamental erotic yearnings of us humans. They have tended to ignore the third or meaning pleasure. Research psychologists have stressed homeostatic pleasure and also meaning–finding but have ignored the second erotic pleasure. Until recently they also ignored conversation pleasure but this has been more than made up for in the last decade or so (Bower, 1977, 1982; Leach, 1974, 1984).

A parent, especially a mother, is frequently a necessary participator in all forms of pleasure, especially when feeding. But certain pleasures are probably essentially found in *solitude*, excretion for instance. And much meaning and mastery pleasure can only be found by a baby or older person when to all intents and purposes he is alone.

Parents as Sources of Displeasure and Distress

Earlier in the chapter we have thought briefly about pain, frustration, overstimulation, cold, hunger, lack of support and sleeplessness as causes of distress. Later we have alluded to understimulation and we will hear more about intellectual and emotional *confusion*.

There are differing views about how parents, mothers particularly may affect the enjoyment or distress of their children. There is by now so much evidence that in general ways parental influences on children are enormous that I am sometimes puzzled why anyone bothers to question it (Ainsworth, 1978; Bowlby, 1969; de Mause, 1974; Leach, 1974; Schaffer, 1977). However there is value in such questioning for we can then argue different points of view; it is by this that we can continually refine our ideas about causes of enhancement and distress. For instance it seems to be becoming clearer that it is often the general *atmosphere* emanating from a parent, given no doubt by subtle cues, that distresses children more than any easily definable specific activities.

There is even more controversy about the *lasting effects* on a child of infantile distress. Babies are certainly less fragile than we used to think especially if periods of distress are made up for later (Ainsworth, 1978; Bowlby, 1969, 1973; Rutter, 1972; Slukin, 1983). But this must not make anyone complacent. These are vital questions, not only for those concerned with child care and custody, but for all of us who are parents or will work with children. Our children's and grand–children's futures depend on how we address these problems.

Some More Theoretical Considerations About the Early Months

The appearance of a new integration, such as, say, smiling or achieving a hand–eye co–ordination, is popularly called a *milestone*. A summary of common milestones is termed a developmental schedule (Sheridan, 1968). Such schedules are commonly used by health visitors, medical officers and paediatricians as useful preliminary checks upon an infant's development. But they are only rough guides since babies develop particular skills differently, both individually and from one culture to another.

We noted earlier that the means by which new integrations come about is presumed to be by an interplay between the maturing of *innately delimited patterns* (usually called reflexes) and *learning from experience*. We know now that infants must learn at an amazing speed. How they do learn has long been the subject of psychological research and debate.

Passive or associative learning is known to be possible *in utero*. For instance, the following classical *conditioning* sequence has been observed in the eighth month of pregnancy. A buzzer sounded close to a mother's stomach evokes no response from the foetus but a vibrator resting on the skin evokes kicking movements. If the buzzer and vibrator are applied together several times then naturally kicking also occurs. If the buzzer now sounds without the vibrator the foetus will still kick in response. Similar conditioning or passive learning of associations takes place after birth. For example, a mother's face is soon conditioned to or associated with feeding as is evident from a baby's preparatory sucking movements. Hundreds of concomitant impressions, associated together, must link to form new meanings for a young infant.

But if all learning were passive, meanings would be over-whelmingly confusing for the infant. With regard to this it has been noted, particularly by Piaget and more recently by many other workers, that a baby is from very early days *active* in being interested in certain things and in ignoring others (Bower, 1977; Bruner, 1976; Piaget, 1953). What is more, he spontaneously *explores* stimuli and hence begins to integrate meaningful structures of ideas and actions of *schemata* for himself. For instance, he may by chance lick his lips, seem to get pleasure from this and try to repeat it. On subsequent days he will repeat the licking with greater sureness, apparently just enjoying the active mastery for its own sake. This repetition of a sensori–motor pattern is called a *circular–reaction* by Piaget. He suggests that by this active exploration and repetition the infant builds up for himself integrated structures or hierarchies of meaningful activity as days and months go by. They are termed hierarchical because new patterns of activity are clearly dependent upon simpler ones mastered previously. For instance, in early weeks a baby is relatively passive in feeding except for the sucking of his mouth and tongue. This in itself becomes more assured through repetition. Then, as mentioned earlier, he fairly quickly associates his mother's face with feeding and actively orients his whole body and manipulates her breast to start efficient sucking. Thus the original sucking is subsumed in a hierarchical way under a wide pattern of activity.

Piaget and his many associates (Boden, 1979; Boyle, 1969; Piaget, 1953, 1972; Phillips, 1975; Vinyk, 1981) have developed a comprehensive theory of intellectual growth delineating how these early sensori–motor *schemata* are slowly transformed, through the stages of childhood, to fully logical, abstract rational thought which is only to be achieved in adolescence. Each new stage arises out of integrations of earlier modes of mental activity and cannot develop without the mastery of earlier stages.

Other psychologists affirm the importance of processes akin to the circular–reaction of Piaget in early learning. Many fiercely criticize him however, not only for the small samples upon which he made his observations, but also because they doubt whether the details of the timing of his sequences is justified by even his own observations, let alone by the findings of other workers. Nevertheless no other group of workers has approached his breadth and depth in the theory of intellectual development. He is, like Freud, essential to a deep understanding of development. On broad issues he is one of the greater 'horizon shifters'.

Psychoanalysts have made little contribution to understanding the details of intellectual development. They have been much more concerned with emotions and the fantastical aspects of imagination.

The Significance of the Idea of the Oral Phase

We noted earlier that Freud and psychoanalysts (Freud, 1905) call these days of infancy the early oral phase. This is because the mouth for breathing and feeding is obviously so important. Functions concerned with feeding must immediately get organized relative to his mother if an infant is to survive. Other parts of the body can and do wait till later months, but the mouth, gut and lungs cannot wait. We have, however, noticed that these early days are concerned not only with the mouth but particularly with the use of the eyes and nose as well. We might better call these days the 'snout' phase.

Early oral patterns certainly seem to persist throughout life.

Consider for instance the social *expression of feeling* or *emotion* (it is a useful convention to reserve the term emotion for the open *expression* or communication of feeling). Much of this is by the shaping of the mouth as well as in the use of other facial muscles, hands, and posture (Argyle, 1980). People often bare their teeth in rage, purse their lips in meanness, drop their jaws and open their mouths in wonderment or innocence and so on. Darwin, incidently, noted all this years ago.

More than this however, our everyday language is full of phrases about oral functioniing. Here are a few: 'I've bitten off more than I can chew', 'He makes me sick', 'She's sweet', 'She's sugary', 'I can't swallow that', 'Trust not a lean and hungry man', 'This will make you choke', or 'I love you so much I could eat you'.

These everyday phrases are not normally about eating or drinking. They are metaphors; they are in the evocative language of oral activity but about other things. They show the psychoanalyst and, I think, anyone else who dwells a moment on the subject, that *oral fantasies* infuse much of adult thinking especially when we are being emotional (Segal, 1985).

Notice too that non–differentiation of self and object is common in such oral fantasies and that it bears distinct similarities to the primitive registration of sameness which we discussed a few sections back. There we noted that sameness seems to precede similarity which involves the discrimination of differences as well as sameness. Self–object difference, like other registrations of difference, is not yet stabilized.

Primitive Emotions, Non-Differentiation, Self and Objects

Turning to emotions generally, our baby certainly soon manifests contentment, excitement, rage and confusion. Perhaps after a few months it can be said that he is joyful or elated and also depressed. But there is *no evidence* for such complex feelings as sadness, hope, pity, concern or revenge. These emotions involve complex structures of intellectual discrimination and feeling some of which may be put into words as: 'I have lost', 'I expect to get but haven't got', 'He, not I has been hurt', and 'I, not he has been hurt'.

From comparing gestures indicative of emotion in infants with those in older children, together with our experimental evidence about the intellect (Bower, 1977, 1982; Piaget, 1953), we have two lines of evidence to suggest that our infants have not yet discriminated, in any stable way, that which belongs to themselves from that which belongs to their bodies and to the outside world. Much intellectual learning must take place before a baby can sort out what belongs to self and what is not–self. This leads us to conclude that *self–object discrimination* is, at most, unstable and rudimentary in the early months of life. If this is so, then, in the experience of a baby, a mother's smile is not meaningfully conceived as being less or more his possession than is a hunger pang. And the sensations of a bowel motion are no more or less outside him than the joggling of his pram. More dramatically, our infant has not yet conceived that his mother's breast, say, does not belong to him. It is no more nor less part of himself than the feelings of his tongue or the distendedness of his stomach. These may be discriminated but this is not the same as being conceived in terms of belonging to the self or not. The signs of early self–object discrimination will be considered in the next chapter. Here I would like to stress only how important this state of non–differentiation seems to be. Its presence has already been stressed when describing the child's early relationship with his mother and their mutual mimicry and synchrony. This early state of non–differentiation is not only very enjoyable to both mother and child, but learning also seems to take place at a very great pace when there is plenty of synchrony. It probably forms a basis of comradeship, paliness and friendliness of all sorts.

On the other hand, when non–differentiation chronically persists into all aspects of childhood and beyond then a person is able neither to think rationally, nor relate to others realistically, nor to adapt to the external world. Such a chronic state of self–object non–differentiation is *psychotic* and is naturally very disturbing for the sufferer and those around him. Incidently, psychosis can in no way be said to be caused by happy non–differentiation between mother and baby. If anything, it may be rooted in an insufficiency of such happiness in infancy.

It seems that transient or partial states of non–differentiation are both normal, in that they are frequent, and also probably essential for deep communication and health at any age.

Non–differentiation is primitive but not necessarily pathological, just as drinking milk is primitive but not necessarily pathological.

Thinkers in the Western world, and psychiatrists in particular, probably undervalue transient non–differentiation. Perhaps this scepticism is a by–product of our deeply ingrained assumption of the unique superiority of Aristotelian logic which assumes subject–object differentiation as Persig (1974) suggests. Certainly in the East the prejudice has for centuries been, if anything, in the other direction. The Buddha, for instance, propounded that unnecessary human suffering originates essentially from *spurious* self–object differentiation. Zen and other Buddhist masters have cogently argued and refined this point of view ever since.

Primitive Classification Later in Life

At a more mundane or everyday level we have already noticed how oral fantasy pervades even adult emotions and feelings. Likewise the crude rudiments of classification which we have referred to several times in terms of registration of sameness also seem to be omnipresent in adult mental life. For instance in *prejudices* voiced in such phrases as: 'They're all the same to me, just wogs', 'Men are all the same, lazy pigs', 'Oh a bad driver, must be a woman'. These thoughts employ mental mechanisms of quite different sorts but they all slip into a *gross perception of sameness about a class* of people when differences are also realistically important.

Many years ago Freud (1900) pointed out how our primitive emotional ideas (or unconscious as he called it) do not obey the rules of ordinary rational logic. Peculiar things seem to happen at these levels of imagining. We have not the space to describe them in detail here, we can only try to begin to unravel a few of these peculiarities in a rather oversimplified way as we go on. Suffice it to say here that very recently, following from where Freud began, the psychoanalyst Matte Blanco (1975) has started to specify what actual peculiarities of logic take over when ideas are emotionally charged.

Matte Blanco's reasoning is very difficult, but putting it simply: emotions must involve thoughts, but they are thoughts where *sameness* of things are registered in gross ways. There is a lack of full discrimination of *differences* of relationship. The latter is, of course, characteristic of rational thought. For instance, when we are in love with someone everything about them tends to be pure loveliness, discrimination of their different aspects falls into the background. A glow of the same loveliness even infuses everything else in our lives at the time. Or when we are depressed, everything tends to feel useless and hopeless even though rationally we know that things can be discriminated more complexly than that. Here again a sameness has taken over.

If it is so that our emotionality is permeated by slips into registering only samenesses when more complex similarities and differences can be seen by others to apply, then naturally, these primitive irrational processes continue on into adult life. However, unlike an infant, with care and good fortune we can not only have the richness of emotionality but also the thoughtfulness of rationality.

FURTHER READING

1 Bower, T. (1982), *Development in Infancy* (Reading and San Francisco: W. H. Freedman). This is a primer of infant development from the research point of view. Includes much of Bower's own original findings.

2 Leach, P. (1974), *Babyhood* (Harmondsworth: Penguin). Full of information of value to both students and parents. Combines scholarship and research findings with friendly advice and conclusions. There are several other books by Leach also worth reading.

3 Lewin, R. (ed.) (1975), *Child Alive* (London: Temple Smith). A very readable summary of research evidence.

4 Roberts, M. and Tamburri, J. (1981), *Child Development 0–5* (London: Holmes McDougall). A systematic course of exercises for those on Child Development Courses. Invaluable for those who wish to be thorough.

5 Robertson, J. and J. (1982), *A Baby in the Family* (Harmondsworth: Penguin).

6 Shaffer, D. and Dunn, J. (1980), *First Years of Life* (New York and Chichester: J. Wiley). A good general primer.

7 Sheridan, M. (1975), *From Birth to Five Years* (Windsor: NFER and Nelson). For 'milestones' particularly.

8 Sylva, K. and Hunt, I. (1979), *Child Development a 1st Course* (Oxford: Basil Blackwell).

CHAPTER 4

The Second Six Months

Sitting Up and Moving About

These months are usually a real delight for everyone. The early days of worry, doubt and broken nights have passed and the days of running about into danger and everything else are yet to come.

During the early months our baby will have been progressively more and more awake during the day, but still with rests which give mother a break. He will probably have stopped his middle of the night feed so his parents will be more rested as well as confident if things have gone more or less well so far. However, remember some quite ordinary babies may start being wakeful about now for some reason of their own, for which we can give no general explanation. This is a time of enormous, easily visible, thrusts in activity and these in turn lead to crucial emotional developments.

At six months he will usually still be fed by breast or bottle but also by spoon and cup. We have already seen that he has begun to co–ordinate his hands with what he sees. At feeding times this often means quite a lot of mess because he usually wants to grab and play with the cup, spoon and plate as well as get as much food into his mouth as possible. His back will be getting stronger so that he can sit up straighter for longer, he will need pillows around him to stop him toppling but by eight months or so he will usually be sitting up unsupported. This is what is real *fun* for him; he can sit up and see all the world around him. His focus is no longer short–sighted so that he can see people and things near and far. He is beginning to interact more generally with people; grinning, waving, loving to wriggle, and athletically cuddle with them.

The second six months is crucial with regard to becoming a tool–user. We noted that by six months old he can reach swiftly and accurately for objects. At this age he uses the whole palm, not yet finger and thumb opposition. He usually drops a thing he is holding if offered another, he can't do much with two things at once. By ten months old he will just about have discovered how to throw things away and then pick them up. Before then there will have been an infuriating time when he delighted in throwing a thing away and shouting for someone to pick it up. Later in the year he will be eagerly relating two things together so that by a year old he will probably, say, retrieve one toy by using another. Here an object is used actively with an *intention*, it is a *tool*, we will say more about this later.

Soon after sitting up unsupported will come some mode of *crawling*, this means self–propelled motility of his whole body. The range of his interest is greatly enhanced. It also means the beginning of the end of safety for him and peace for his mother. He can be a danger to precious things and to himself. Breakables can't be left on low tables or floor; but, more important, electric plugs, appliances and stairs must be kept safe. He is beginning to have *his active effect* on the world as well as it on him. By a year old he will probably be able to pull himself to standing and even walk hesitantly holding on to chairs and things. All these are marvellous for a child, chortles of pride and triumph make this obvious. Most gratifying of all for many parents is that he will be more and more actively interested in other human beings. He tends to be puzzled and tentative with strangers, reacting quite differently to them compared with his parents. But given time he will readily turn away from his parents, at least for short periods, and enjoy playing with others. Towards the end of the year he will probably show definite signs of self–conscious *coyness* and a wish to *show off*. He will also not simply *mimic* others in the minute ways described in the last chapter, but will, again with apparent self–consciousness, *intentionally* play at imitating them. Playing with sounds is enjoyed by babbling and repeating intonations to himself which are seemingly meaningless. These sounds will later be refined to become recognizable imitations. Probably by a little older than a year, the first recognizable words for objects will be developed. (Leach, 1974, 1985).

The Need for Moderate-Newness

As we mentioned in the last chapter, a baby in learning to cope with the world omnivorously seeks out and registers the multitudes of relations of things presented to him. He memorizes the repetitions of things that are the same as before and also those that are new and different. He is eagerly learning about the things in space around him and naturally needs interesting objects to play with. They don't need to be expensive toys, a collection of safe unbreakables from around the house is quite as good. Totally new things startle him and too many of them are very confusing, but nothing but one or two very familiar toys are of no interest to explore. They get boring, there is too much sameness and not enough difference. Under stimulation is a really serious deprivation for an infant (Leach, 1974) though many parents do not seem to realize this. No doubt in old village cultures there was always something new lying about to pick up or to see from mother's back. But shut in modern apartments, often with a mother who is equally a prisoner, a baby may be shielded from traffic and electricity but also cut off from natural everyday stimulation. This then has to be provided by parents' forethought, which is not easy when they feel themselves to be prisoners too.

Weaning, the Later Oral Phase

We have already mentioned our baby feeding by spoon and cup, this means that he has been eating solids as well as suckling from breast or bottle. Into the second half of the year his first teeth will be appearing, this means the beginning of biting, grinding and chewing. These are quite different mouth actions from sucking, they herald the beginning of *biological independence* from feeding upon his mother. Although a baby can suck from the breast even with teeth for a long time to come, it does mean that a mother is reminded that weaning will soon be appropriate. She will painfully be reminded of this if he gets to biting her much.

It is worth noticing that weaning is a model, as it were for many partings that need to be worked out between child and mother as he grows up and away from her. For many mothers breast feeding

49

is a sheer delight, deeply, erotically, enjoyable for mother as well as child. It seems to be a very sad thing to end it and quite a few mothers put it off as long as possible. Certainly too early weaning can be an awful experience of loss for a baby. On the other hand there does seem to be an optimal time to slowly cut down on suckling. By and large it seems that it is best, in Western culture at least, to think of having it complete by about a year old. Going on after that often means that mother and child are stuck together emotionally, when other mothers and children are more separated, and it is very difficult to break the habit.

It is useful to think of mothering not just as providing for a child's immediate needs but for his future. With this in mind weaning can be seen as positively helping a child to enjoy the next stage of his life, it is then recognized as giving something new to a child as well as taking something away. Remember, difficulties in weaning are just as often because of reluctance to give up the pleasure by the mother as they are due to a real need to continue by her baby. Sleeping difficulties at this age may be due to a non resolution of weaning (Kitzinger, 1979).

Mother, Father and Baby

The food and playthings we have just discussed are vital provisions of a physical kind. But what parents provide covers not only these concrete things but also the way they behave and are *emotional to each other* and their baby. We saw in the last chapter how valuable the 'conversations' between mother and child are and how the style of her handling and her rhythms affect her child, her atmosphere is important (Ainsworth, 1978; Clulow, 1982; Krames, Pliner & Alloway, 1975; Leach, 1974, 1983; Schaffer, 1977).

This atmosphere is likely to be dependent upon the back–up a mother gets particularly from her husband. But as time goes by he, as father, is likely to come more and more into our baby's eye view. In the early months, if he has been a keen father, he will have been eager not only to help his wife but also to bath, change, handle, cuddle and play with his child. It is stressed these days how important it is for a mother and baby to have a cuddle immediately after birth. It is probably nearly as important for a father to be in at the beginning so that he sympathizes, gets in synchrony with, and identifies with his child from the start. He is then more likely to be confident and easy as a father. What is more, knowing his baby's routines means that he can free his wife to have a break. He will never be able to have a baby in the womb himself nor to breast–feed. He is excluded absolutely from these, but he can do most other things that a mother can do.

As well as doing things that are similar to a mother there is one other thing that he provides directly to his baby all on his own: he is *different* from mother. His muscularity, movements, style and sounds all provide a moderate newness for the baby. This becomes increasingly important as a baby gets more mobile; then a father, who very often enjoys his muscular athleticism, comes into his own. When a father really gets involved with his children as babies it is great fun for everyone, I have never known a father regretting it (Biller, 1974; Cath, 1982; Lamb, 1981; Parke, 1981; Rapoport, 1979).

Into the second half of the year we see that a child is not only beginning to relate two physical objects together, as we saw with the coming of tool–use; but he also begins to relate *two people* together. We can see him looking from one parent to another, grinning at both, getting cross if they spend too much time talking to each other and so on. His range of emotional gestures and mobility is still limited so his capacity to get into their business, to *affect them intentionally*, is still small, but there are signs that it is beginning. We will hear more of this in later chapters.

Digressing for a moment from the intimate family to a wider social view. Our frame of reference is a western style nuclear family; the parents of babies from Africa and Asia particularly will find some of the habits mentioned here strange. Not only may the traditional styles of husband and wife relationship be different, as we all know, but also physical movement, styles and practices can be very different from one area to another. For instance, many African mothers and babies enjoy gymnastic performances which quite frighten unwary, unathletic Europeans. Maybe this early physical enjoyment accounts, in part for the natural grace of African dancing and athletic movement, compared to their hesitantly formalized European counterparts (Yarrow, 1977; Wagner, 1982).

However, whatever the style the more both parents let

themselves enjoy their child the more will he reward them by his enjoyment of them.

Good and Bad Experiences

Returning to the intimate experiences of our individual baby, we have, on several occasions, shown how important is the pleasure a baby gets from his mother. At the same time he must also inevitably be distressed by her on many occasions. Some mothers distress their babies more than others; but, the world being what it is, all mothers must do this some of the time.

It is obviously of prime importance for the *survival* of any animal *to discriminate* what is *good* for it and what is *bad*. This is essential from the earliest days of life and the human infant is no exception to such sensitivity. Dwelling upon this vital issue the psychoanalyst Melanie Klein (1948; Segal, 1973) introduced the distinction of a *good object* and *bad object*. She particularly stressed an infant's experience of his mother in these two ways. It is a funny sort of way, perhaps, to talk about a mother, but it brings out some most useful ideas, not only about babies' fantasies but also when thinking about later life.

Everyday language tends to confine the word object to inanimate things. For quite good reasons psychoanalysts, when talking about primitive feelings and imagination also, use the word object to refer to all aspects of humans, animate and inanimate, as objects. The point is, that, in our most primitive emotional imaginations, as we began to see in the last chapter, fine distinctions such as between animate and inanimate are not yet made, or at least are instable.

We noted how the difference between self and not self is probably nearly non–existent, not only in infancy but also in our *primitive emotions* later in life. So too is the difference between being alive or not. For example, we can mix up animate and inanimate in everyday adult life when, say, we are late for an appointment and the bike or car decides not to work. The more hurried we are the more likely we are to shout at the machine, even kick it, as if it was being an intentionally obstinate person. Equally people can readily be treated as things. With this in mind the generalized use of the word object when referring to emotional happenings is not as silly as it sounds at first.

Returning to a good or bad mother; when she is experienced as *enhancing* a child's pleasure (homeostatic, erotic, social or meaningful) she is obviously welcomed with joy, in which case she is 'good'. When, on the other hand, she is experienced as arousing distress she is being 'bad'. What is more, babies tend to react in *all–or–nothing* ways, so experiences are either totally good or totally bad. This is called *splitting* into good and bad, it is a primitive way of preventing confusion. There is, thus perhaps, a proneness to experience things, mothers in particular, as infinite goodness or badness. So the infant's biological need to discriminate good and bad can evoke fantastic reactions which have consequences in learning and adaptation for years to come.

These notions of good and bad only refer to the *immediate* experience of a child. They do not refer to what is good for him in the long run. Quite clearly, not only is it impossible for even the most devoted mother to be good in this sense all the time, but also, what is felt as bad for a baby in his immediate experience may well be really good for him in the long run.

Another psychoanalyst Spitz (1965) has stressed a similar distinction between good and bad in the mother–child relation, but in the opposite direction from that of Klein. Klein pointed out the importance of a *baby* splitting and hence feeling his mother to be either all good or all bad. Spitz, on the other hand, has pointed out the importance of the *mother* feeling her baby to be good or bad. These states, he suggests, are signalled by 'yes' and 'no' gestures from a mother to her baby. 'Yes' gestures, sometimes consciously intended but often quite automatic, are made by the mother when she is enjoying what her baby is doing. They help to affirm to the baby that he is good for his mother, and hence, since mother is most of his world, that he is a good person, worth being alive. 'No' gestures, on the other hand, negate what a child is doing and probably tend to instil in a child a sense of being a bad person who is not worth having around. We can detect here the possible roots of *guilt* which will be discussed in the next chapter. What is more, when a mother makes 'no' gestures she rebuffs the child who is then likely to become, if only transiently, distressed. Hence, mother and child naturally feel mentally bad for each other. Happy lovingness is 'dead' for the time being. Remember, we are not yet talking about discipline, that must wait till the next six months of life. We are talking more about a mother being grudging to her baby. Every exhausted mother will be grudging a bit of the time, what is being stressed here is that babies probably notice this quite as much as they do their mother's love.

In summary, when a mother evokes distress, and all that it entails, she is experienced as bad. As a mother must inevitably be at best both bad and good at times, then she is the child's first enemy as well as his first lover. Furthermore, the patterning of these good and bad experiences seems often to set the style of child–parent communication and hence, perhaps, for the climate of much of his later life. They point up some important aspects of the *atmosphere* between mother and child (Leach, 1974).

Impulse and Fantasy: Before Steady Differentiation of Self and External World

We are slowly getting to the point where our baby begins to recognize steadily that he is a being separate from external reality: the outside world of things, spaces and people. This differentiation seems to have begun to happen clearly, for normal children at least by some time in the second six months. It is a slow process; we have noted that there seem to be innate mechanisms, those used in distance perception for instance, which play a part in sensing the external world. These must be present from the beginning.

On the other hand we also have noted many signs that self and external world is not yet being clearly structured. We have noted in the previous chapter the omnipresent 'mistakes' in localization, blurredness of vision at a distance, also the inability to manipulate separate objects relative to each other and so on. There is further evidence about non–differentiation from the way babies cry: very young ones just seem to howl reflexly; but somewhere about six months they begin to 'call' in their crying. They seem to have become aware that, say, mother is an object 'out there' who has gone and can be recalled.

There are lots of bits of evidence like this, none conclusive in themselves, of what it is like for a baby, but together they point to the inferences that we can summarize as follows. A very young baby does a lot of differentiating but he is not sure of either the constancy of external objects nor of the difference between himself, his body and these externals.

When we try to imagine ourselves into this non–differentiated state of the young infant, as we did a little in the last chapter, we logical adults will probably feel it is a bit mad and get irritated with the exercise. But do remember that unless we are willing to imagine a bit madly we are not likely to understand infants imagination, nor our own at primitive emotional levels. So let yourself go, see whether you can wonder yourself being situated in your own thumb, or in the nose of the person next door! Try imagining that someone else has become you or that Mount Everest is really a person! These are quite mad ideas to our sensible conscious selves, they are certainly what some mad people voice, but something like them happens to all of us when our emotions take over: they occur every night in our dreams too. So it does behove us to exercise this madness; *and* to get back to being sensible.

Returning to our baby. It looks as if his waking imagining must be very much concerned with *bodily happenings* and things *immediately visible and touchable* because these are all the data his narrow experience has to use. We can assume also that his bodily happenings are felt no more nor less part of himself than outside things. Thus his mother can be part of himself and he part of her.

What is more, as his life is centred on mouth, nose, hand and eyes, we would expect much fantasy to arise around these zones. It is likely to be full of tasting, gulping, spitting, burping and, after the first teeth, by biting, chewing, grinding, puncturing, tearing and so on. It is also likely that he imagines passively being swallowed, chewed and spat out; remember his sense of himself, if present at all, could at times be felt as able to go anywhere and have anything done to it. When contented, a baby's fantasy is probably experienced as mild and pleasant. But *in distress* it is likely to be violent, explosive, in bits, and discordant. Grinding, slashing, engulfing, ejecting, tearing, exploding would then be most likely fantasies. What is more at this age every experience must tend to have an urgent overwhelming immediacy. Things will feel limitless and inconstant.

Just as space is not yet recognized as constant, so is it unlikely that *time* is. Things that happen may then happen *infinitely*. This is because an *end* to an experience is not yet clearly anticipated.

In this light we can speculate that when a baby feels that something is bad it is infinitely bad, and when good, it is infinitely marvellous. We ourselves can sometimes get glimmerings of this sort of emotion about our own mothers if we cast our minds back into the distant past. We can then easily have an image of her as a formless evil creature, nearer infinite devil–witch than human; or as a wondrous being like a goddess. This is likely to be how a baby imagines when being emotional (Matte Blanco, 1975; Rayner, 1983).

Loss, Separation Anxiety and Deprivation

We can begin to see that, in building up the idea of himself as separate in a world of space and things, our baby must do a lot of throwing away objects, losing them and then finding them again. As we saw earlier he is doing this in his second six months just at the time he is giving evidence of making the crucial self–other differentiation.

But at the same time as discovering about toys coming and going so too he seems to be finding out coherently that *people come and go*. We noted that he has begun to *call* his parents. With this *separation anxiety* appears.

In the early months emotional distress is obviously present but not specific *anxiety*, which involves imagined *anticipation of future events*. In the second half of the year a baby will fret when his mother puts on her coat to go out, this is separation anxiety. And often, if she stays away for more than a few minutes leaving him in a strange place he is likely to become extremely distressed.

At about the same time parents are deeply distinguished from unknown people: *stranger anxiety*, begins to appear (Ainsworth, 1978; Bowlby, 1969; Decarie, 1974; Krames, Vol. III, 1975). Both point to the beginnings of *anticipation of loss* of loved ones. This brings us to a very important question indeed. What are the real effects of loss for a young baby? Are fears about loss justified? It has been known for several decades that little children, particularly under the age of three, who lose their mothers for any length of time and are left in a strange environment, first become extremely distraught, then after a few days, quieten down and become 'better behaved'. If the loss continues they seem to 'give up', become apathetic, manifesting most of the symptoms of profound *depression*. On continuing separation they may seem to brighten up but the vitality is shallow.

Although sensitive parents have been intuitively aware of this reaction to loss for a very long time, it is remarkable how many cultural practices have ignored it, not only in this country but throughout the world.

Forcefully enlarging the work of Spitz (1965), Bowlby (1951, 1973) and his colleague Robertson in films brought this question towards formal recognition, by pointing out this syndrome when children were sent to hospital or to children's homes. We owe them a great debt for their clarity and persistence in making a vital change in our culture and care of young children.

Since then there has been a great deal of research and argument about how much a young child can be separated from his familiar mother without serious damage (Clarke, 1975). This has been summarized by Rutter (1972). In general, his conclusions can be briefly stated as follows. If a young child is separated from a familiar mother and placed in an unfamiliar and, to him, chaotic environment then the picture of distress passing to depression, as described by Bowlby, does take place with full severity. If, however, a child is separated from his mother but placed in a familiar environment with a relative or foster person, with whom he is also familiar, then there is much less sign of distress and probably few, if any, serious long–term effects on development. This contention is movingly argued and demonstrated in the well–known films by the Robertsons (see film list at end of this book).

A similar, but not identical, problem with regard to infant development was raised some time before Bowlby by Spitz (1965). This is the question not just of the loss of a good mother but of *deprivation* of mothering. Spitz observed that children in large institutions, where they were well nourished and clean but left chronically devoid of consistent stimulation, became extremely *retarded*, both intellectually and emotionally. Furthermore, they were very prone to infections so that mortality rates were far higher than less sanitary but more friendly environments. Thus deprivation of affectionate stimulation was seen to have not only psychological consequences but also serious somatic ones.

More recently, health visitors and social workers have inform-ally noted how frequently these deprivations of consistent stimulation and conversation can occur even for a baby at home with his own mother. Deprivation is not confined to infants in hospitals or large institutions. This concept has been extended to that of the *cycle of deprivation*, which points out that parents who were deprived in their childhood probably tend to repeat the pattern with their own children (Kelmer-Pringle, 1974).

It is worth recognizing at this point that the overall concept of deprivation can be a crude classification. We ought, perhaps, not to rest content with it but rather to specify in what particular functions a child seems to be deprived. We have already noted deprivation in availability of playthings and stimulation. Some children are deprived of nutrition, others of affection, others have plenty of affection but little quiet consistency of stimulation, and yet others may be deprived of variety while experiencing plenty of affection and consistency. Proneness to crude classification by workers can be most unjust to mothers who are doing their best but failing, probably unconsciously, only in certain aspects of their responsibilities. There is a certain brutality in negating all a mother's worth instead of affirming some aspects while critically drawing attention to others where failure is really occurring.

To many of you, these arguments about loss and deprivation will sound like obvious old hat. The findings have certainly been with us for a long time now, but it is dispiriting to recognize how easily people, who ought to know better, forget these things. Thus professionals still can be blind to their own children's lonely plight as they absorb themselves outside the home. And politicians on both left and right can destroy well thought out provisions for children in the cause of financial stringency or in the service of a political ideology. Profoundly valuable knowledge which affects people's lives can be destroyed by thoughtless people as easily as anything else.

The Psychological Birth of the Self and the Sense of External Reality

Let us now return to our individual baby and work out some theoretical ideas about him. We have, for a good many paragraphs now, been circling round various facets of the discrimination of self and not–self, which goes with discrimination of *internal* and *external reality*. Mahler (1975) has called this the *psychological birth of an infant*.

The fate of this function of discriminating what is internal to oneself and what is external is momentous for every one of us. Its advent begins to allow a child to grow into a separate intelligent sociable being. We have also noted that the differentiation of internal and external fluctuates for even the most normal person. However we have already seen that when someone's sense of self–differentiation becomes chronically distorted, compared to the way the rest of us think, then we get a feeling that he is *mad*. Thus a person who feels all the world belongs to him, or that he is responsible for an earthquake in China, or who feels he is being controlled by agents of an alien power, has clearly got the senses of himself and external reality distorted. He is fairly quickly thought to be *deluded* or *psychotic*.

Man's sense of himself has, of course, been an abiding concern of civilized mankind in both East and West for thousands of years. The religious and philosophical literature on the subject is vast and quite beyond me to summarize. In past decades, perhaps the most vigorous analysers of experiences of self have been existential thinkers (May, 1967; Tillich, 1952). Academic psychologists have on the whole bypassed the question, a notable exception is G. A. Kelly (1963) whose theory of *personal constructs* has given birth to a well–known school of research. Again, until very recently, psychiatrists have had little to say on the subject. With the exception of Jung, psychotherapists have also only recently addressed themselves to the question. It has now become of very active interest (Winnicott, 1965; Kohut, 1971; Khan, 1974). Mahler (1975) and her associates have perhaps been the most systematic psychoanalytic investigators of the development of self–differentiation in infancy and early childhood. Led probably by the original thinking of G. H. Meade (1932), sociologists have produced searching work on the social contexts of self–feeling (e.g. Goffman, 1959). A visitor from another planet would probably be struck by the present intellectual ferment on this subject. There will be no answers to the question of self in this book, but let us briefly think about one or two facets which may provoke further questions.

The necessary substrate of each person's self–feeling must be that he is able to be aware. We do not know when an infant's awareness begins: it seems to be wakefully present at least from the earliest days after birth and probably even before then (Verney, 1982). In the past it has been usual to equate awareness with wakefulness, but our understanding of *dreaming* (Freud, 1900) indicates that a form of awareness continues in sleep. Recent physiological studies indicate that this dreaming occurs in infancy, even *in utero*, as well as in adult life (Wolman, 1979). Experimental studies of subliminal–perception likewise stress the presence of subconscious awareness (Kline, 1972).

But awareness is not the same as self–awareness which is our present concern. The pronouncement of Joshu, a Chinese Zen master of a thousand years ago, was simply 'no–thing' when questioned about the nature of the self. It seems useful to follow his advice and stop thinking of the self as a concrete entity. Rather let us recognize it as one side of an *act of discrimination*. We cannot meaningfully talk about 'the self' as if it were an entity in isolation but only in relation to not–self experiences.

One facet of not–self feeling is the experience of *external reality*. This seems to be stabilizing in the second six months of life. We have already noted our baby playing at throwing things away and getting them back or looking for lost objects. He seems to be discriminating that 'things are real out there'. Obviously ideas of external reality are dependent upon patterned information from the external sense organs: sight, hearing, touch and movement. But ideas of *constancy* are also involved, things can disappear from sensory contact yet be conceived as still present somewhere in order to return later. Stable memory and imagination are important in this (Boyle, 1969; Phillips, 1975; Piaget, 1953).

Emotionally, external constancy can be seen to entail two sorts of experience: first, awareness of *loss* and, secondly, expectation or hope of *return*. For these to occur, and hence for a baby, or an adult, to feel something as real, he must recognize it to be out of the control of his *whim*. For instance, you feel the chair you are sitting on is real and constant not only because you can touch and see it, but also because you intuitively know it can be taken away, it is not subject to your every whim. If you wished, perhaps you could move or smash it, but you know you could not eat it or magically turn it into a butterfly.

This inability to control must entail *frustration*, especially for a baby. And we know that this evokes distress and *rage*. Thus the development of a sense of external reality inevitably involves feelings of rage, which means destructive aggression, as well as the experience of loss. Furthermore, if the sense of self is the counter part of external reality then the feeling of *oneself* as a separate individual is also born out of frustration, distress and rage. Without aggressiveness there is thus no self. This rather solemn conclusion is not, I think, widely discussed in the literature: it is however, explicitly stated by Freud (1915a), Winnicott (1958) and Rochlin (1973).

The experience of reality cannot simply be one of frustration and distress. Enjoyment and satisfaction also really happen for both infants and ourselves. We know from work on deprivation that the most frustrated babies are certainly not likely to become the most realistic children. On the other hand, it is common knowledge that a child who is exclusively cosseted and pampered to satisfy his every whim becomes a distressing tyrant to himself as well as others. Clearly a baby must develop his differentiation of self from external reality by a tolerable rhythm of satisfaction and frustration. In a mixture of Spitz's and Mahler's phrases the pattern of a mother's 'yes's' and 'no's' provides the background for his psychological birth. Without steady satisfactions and enjoyment from his parents, a baby cannot develop those hopeful ideas of future pleasure that make painful frustrations bearable. When distress becomes chronic, then it seems that a child's mind resorts to a variety of pathological, defensive tricks of the imagination where self and reality discriminations are distorted if not disabled. There will be more of this later.

What we have discussed here are the first dim, vague but deep beginnings of an underlying sense of external reality. We are not yet talking about the clearly structured and precise thinking which we call rational thought, only its origins. The evidence, particularly that from Piaget (1929, 1953), suggests that realistic thinking only slowly develops throughout childhood and after; it is the product of imagination or fantasy being tested and refined by immovable experiences. A child, if healthy, continuously creates hypotheses about the world from his imagination. These are modified, and also distorted, by what he is told, they are also tested by what he experiences for himself.

Sociologists have pointed out how the sharing of a model of reality with others of the same culture gives its individuals a conviction of the truth of this model, even though members of other cultures can see its fallacies. This being so, we must recognize that no representation of external reality, even a formally tested scientific theory, is a final truth. Reasoning doubt must be a hallmark of sane intelligence. Provisional conviction about a view of the world is necessary for decisive action. But a person holding a conviction of the final truth of his ideas must be deluded: unfortunately there are plenty of such people around who have power and influence.

In summary: as the senses of self and external reality seem to be complementary then, not only external reality, but also self–feeling must arise from the combined experiences of enjoyment and frustration. The part played here by frustration may seem strange, and it must be admitted that the evidence from infancy is little researched. But introspection supports the contention that something of a sense of *loss*, *aloneness*, and the ability to bear *loneliness* is intrinsic to sure self–feeling and integrity (there is further discussion of this in Chapter 13).

Infantile Mental Events Compared with Psychotic Processes

It will be getting more plain how tenuous, and fluctuating the sense of self and external reality can be. The position and size of our selves can do extraordinary things. Take, for instance, the idea of the effect of self on others. Here the sense of *power* can easily reach delusional grandiosity. And unfortunately, when a person goes around believing in his own power others are often only too eager to believe him and become followers so that his deluded power becomes socially real, albeit mad.

At the other end of the scale it is easy enough to delude ourselves that we have no effect on others at all. We could call this a delusion of impotence which is common enough when we need to feel guiltlessly innocent and wish to be led. The grandiose and innocent can then form an unholy, dangerous collusion if they get together socially.

In fact the, often unconscious, manipulation of the size and position of the self relative to things in the outside world is one of the commonest mental tricks, or defences, against really unpleasant feelings. For instance, we may momentarily feel ourselves to be intensely bad about something we have done and then instantaneously say to ourself, 'It's not me that's bad it's someone else'. This is *projection*, here something originally felt as within the self is put *outside*. In chronic delusional form it is called, of course, paranoia and must be distinguished from realistic suspicion which is necessary for survival. Its unquestioned action is extremely dangerous. If I was asked what was the most dangerous thing in the world I would not say armed forces or missiles, they are simply implements, I would say 'self-indulgent paranoia in political guise'. Remember here to make the vital distinction between realistic suspicion about a threat and the creation of imagined threats which is paranoia.

In *Melancholia* the 'shape' of the self is rather different, the individual is preoccupied with loss and deadness. Good, enjoyable things are felt to be lost to the self. There is nothing but giving up in *despair*. The self is more or less felt to be so useless and empty of virtue as to be dead. In delusional proportions, of course, this is psychotic melancholia, but at some time or another almost everyone experiences something of this useless deadness, giving up and despair in common *depression*.

Mania on the other hand endows the self with great liveliness, effectiveness and power. Again in delusional form this is mad but manic feelings are common enough in ordinary people. There are also processes characteristic of *schizoid* conditions. These are: being in bits, feelings of catastrophic explosions, of the self disintegrating, of not being in one's own body, and of self–not–self confusion. The characteristic schizoid defense of withdrawal from love of, and interest in, the outside world can proceed into chronic *autistic* fantasizing which, being utterly ideosyncratic, is bizarre to the outside observer.

We can also infer that; since the differentiation of self is critically developing in infancy, then these weird, seemingly mad imaginings of our minds can be present in rudimentary form at that early time. For instance, all the signs of deep depressions are visible in some infants; so also is withdrawn, autistic and other schizoid phenomena. If we listen to slightly older children who can talk we can hear them expressing many of the vicissitudes we have discussed.

But remember: only when he becomes chronically deluded and gives up testing reality on vital issues can someone be truly called mad.

Defence Mechanisms

If it is problematical for an adult to keep sane and content how does the more vulnerable infant manage it? It is obvious that this is not within the baby's power alone. He remains sane largely by other agencies, that is by having loving, sympathetic and consistent adults around him who provide the conditions of moderate newness of stimulation.

But we can also detect the rudiments of *defence mechanisms* which *reduce immediate distress*. Thumb–sucking, auto–erotism and then withdrawal into perhaps transient autistic fantasy is common enough.

Another defence, splitting, has already been mentioned. It concerns good and bad feelings and their 'all–or–nothing' quality. It is easy to observe how a baby or young child has a short time span of attention for one activity. He gets deeply absorbed in a thing, then suddenly 'switches off' it to become deeply engrossed in another. So engrossed is he that it is as if the previous thing is of no importance at all. This switching is very frequently used to recover from distress. One moment a baby will be in a paroxysm of rage, kicking and screaming at his mother, then, if he is picked up and cuddled, he switches, the rage has gone, he is all smiles. 'Bad' mother has suddenly turned into 'good' mother again. The bad experience seems to be forgotten.

This underlying proneness to dichotomize is omnipresent in adult emotional reactions as well as in infancy. Sociologists and anthropologists (such as Lévi-Strauss, 1966) emphasize the dichotomization underlying many cultural practices. For instance, many supernatural beliefs are deeply dichotomous; it has been pointed out, for instance, that even monotheism is really a duotheism or dichotomy between God and the Devil.

At the individual level psychoanalysts refer to this dichotomization as *splitting* (Segal, 1973). As already noted, this refers to experiencing something as absolutely good or absolutely bad. In the process of splitting, when a thing is felt as good, the distressing, bad memories or experiences are *denied*. This is just like the baby who seemed to deny his unpleasant experience of his mother immediately after she had picked him up, cuddled him and he was all smiles.

Splitting seems to act not only as a defence to blot out distress or bad experiences, it also saves an immature child from helpless confusion. If he remains aware of distress and pleasure coming from the same source he is utterly mixed up. Structuring his awareness by splitting allows him to remain coherent to himself, if in an inadequate way. It seems to take much experience and assimilation for a child to develop a coherent sense at any one time of his mother, say, as potentially both good and bad, distressing and enjoyable.

What I have just described is defensive splitting; like all defences it may be useful for equanimity. It cannot by itself be categorized as pathological. Only when it is used chronically and in some way prevents further development can it be so regarded. This would be the case, for example, when a child splits so that he keeps the image of his mother as totally good, whereas other people are felt as evil and he fearfully refuses to have anything to do with them.

Another distinct form of splitting occurs in the actual experience of extreme distress, this is the shattering of the feeling of self and objects into 'bits'. This can hardly be said to be a defensive manoeuvre to ward off distress, it is part of distress itself. However in psychosis individuals do seem to identify themselves with these shattered bits of experience in bizarre ways in frantic efforts to feel some certainty about themselves. Here is a very mild instance of this. After writing the first edition of this book, one night I felt myself to be its pages. It was most unpleasant, but fortunately my common sense ego was also still working and I felt first horrified and then slightly amused. If I had become chronically convinced that I was a book, well . . . . that is madness.

FURTHER READING

1 Bowlby, J. (1953), *Child Care and the Growth of Love* (Harmondsworth: Penguin). An abridged version of his original work on separation.

2 Chodorow, N. (1978), *The Reproduction of Mothering* (Berkeley: University of California Press). Freudian view of the sociology of mothering.

3 Fraiberg, S. (1977), *Every Child's Birthright: in Defence of Mothering* (New York: Basic Books). As it says, warning against devaluation of mothering by a writer of world repute.

4 Leach, P. (1974), *Babyhood* (Harmondsworth: Penguin). Essential further general reading.

5 Oakley, A. (1980), *Woman Confined* (Oxford: Martin Robertson). Another feminist approach to mothering.

6 Roberts, M. and Tamburri, J. (1981), *Child Development 0–5* (London: Holmes McDougall). A worked-out course of observation.

7 Rutter, M. (1972), *Maternal Deprivation Reassessed* (Harmondsworth: Penguin). Just what the title describes.

8 Stafford Clark, D. and Smith, A. (1983), *Psychiatry for Students* (London: Allen Lane). For elementary descriptions of psychotic states.

9 Wagner, D. A. and Stevenson, H. W. (1982), *Cultural Perspectives on Child Development* (San Francisco: W. Freeman & Co.).

CHAPTER 5

One to Two Years Old

The Year of Toddling Movement

At the beginning of his second year our child may be just about taking his first floundering steps: by the end of that year he will usually be moving with assurance. At the beginning there is an anxious pleasure in discovering how to get about. Towards the end, and on into the third year, he shows a sheer joy in progression and the discoveries that come with it. A young child's waking life seems all movement. He absorbs himself in finding new things and then manipulating them, thus discovering what can be done with them, what they will fit into, what noise they make, and so on. *Play is sensory and manipulative.* For instance, sand is ladled from one cup to another and run through the hands, water is splashed about, not only to see what happens to it, but also for its feel. 'Tools' are used to prod, push, pull and beat other things. He will be feeding himself with a spoon, and of course using this to ladle food into other places as well as his mouth, partly through clumsiness but also, to discover the relation of food to other things as well as to himself. The first primitive scribbles will be attempted, perhaps not to represent things, but for the pleasure of movement and to see the marks. Although he cannot draw representational pictures, he will recognize photographs or pictures of familiar objects. Probably by the end of his second year he will be trying at least to be clean and dry during the day (for milestones, see Sheridan, 1975 and Leach, 1974).

He will be understanding words and increasing his spoken vocabulary almost daily. At the beginning of the year he may be using a few approximations. By the end of it he will probably have refined a vocabulary of many words, all clearly recognizable and articulated into simple sentences.

Trust in Self

Before the age of one, a child's mother *is* his intelligence as far as anything to do with moving about is concerned. It is for her to think out and decide where he should be. But when he begins to be a toddler, the job of mothering is to begin to *give up physical control*, and slowly help him realistically and safely to *trust* himself, but also to refine her verbal controls. In one sense her despotism begins to give way to the rudiments of democracy.

This growing trust in himself (Erikson, 1963) is never entirely smooth. He may try too much, fall over and burst into tears, not only because he is hurt but because confidence in himself is dashed. Little children at this age are very ambitious to do what grown–ups do, fears of frustration are very common.

By his explorations our child must be developing a representation of his self as someone who can do some things and not others. We can make such inferences when we see him quite happy about clambering upstairs but, with trepidation on his face, will refuse to go down, having tumbled on one or two occasions.

Erikson has stressed the importance of developing *trust in the environment*, particularly parents, during the *first* year of life as a prerequisite of trust in self. Now we can consider the relation of one to the other. Where a child feels at ease and trusts his environment, he is free to explore spontaneously and develop his own skills, hence trust himself. He will also find it easy to learn from such an environment, to imitate his parents, for instance, and hence be helped to develop the skills that lie behind trust in himself.

Play and Beginning to Form General Ideas or Pre–Concepts

We have already noted that a little toddler's play is certainly still sensory and manipulative. Towards the end of the second year, however, the birth of *active make–belief* will be apparent. When we look at this more carefully in the next chapter we will see that make–belief seems to be rooted in *imitation* (Phillips, 1975; Piaget, 1951). We have seen how fundamental this was in the early months. Now it is still vital in learning but with age takes on a more intentional quality. Active choice is becoming apparent. A toddler tries and tries again to do the same as his elders in finding the means to manipulate the things around him.

One of the results of repeated manipulation is to discover how many things can be experienced and treated as *the same* or, when more sophisticated, as similar. It is the beginning of forming generalized ideas, these too are constituents of make–belief as we shall see. By the end of the year a child clearly understands the meaning of such simple general ideas as in, out, up, down, heavy, light, big, small, few, many, water, wet, loud, soft and many others.

Such ideas cannot yet be called proper concepts, it will be a long time before, say, the concept of heaviness will have the constancy of application necessary for really logical use. There are many ways in which these early pre–concepts differ from what we know in adult logic, suffice it here to note that *primitive classification* often tends to be *wider* than is appropriate. For instance any furry creature may be called a 'doggie' and on the next day a 'pussy'. Only later will the similar animals be more carefully clarified with appropriate attributes, into 'doggie, bow-wow' and 'pussy, miaow' (Leach, 1974).

Note again here that, as the *limits* of classes must be still vague for children, so puzzlements must infuse many things. For instance, when somebody is angry they roar, so is everything that roars angry? Are aeroplanes angry, are train engines too? This is the sort of world of unstable mysterious limits the toddler lives in. As we noted in the last chapter, because limits are vague and unstable there is probably a tendency to 'infinitize'. This seems to have its effect on how little children tend to *feel in extremes*: where infinity reigns one tends to be very, very frightened, or in ecstasies of delight. More of this in the next chapter.

Communication and Language

You will remember how the infant–mother mimicry of body movements in the first weeks of life, extended to communication by gesture and pre–word 'talking' in later months. The growth of understanding the use of *words* and then the formation of these into *grammar* naturally follows. It is the subject of very active research at present (Bower, 1977; Britton, 1985; Lewin, 1975).

Let us first generally recognize that words are signs or *symbols*. Symbols are things which are created to stand for something else. For example words are sounds which have been created by convention to stand for things, actions or relationships. On the whole they do *not* resemble the things or actions for which they stand. A few words do however have this resemblance, for instance 'cuckoo', 'curlew', 'bang', 'plop', 'hiss' and 'piss'; this is onomatopeia. When we turn to non–verbal symbols, visual ones particularly, we see that resemblance is very common. In play material and dreams particularly, symbols do often have the same form as the thing signified. For instance a stick as played with is a similar shape to the gun it represents.

Anyone acquainted with toddlers will know that learning to *understand* words passively, as signs for things develops before learning to use them actively in *talking* recognizably. The active use of a word means that a phase of classification has been completed in our child's mind. Thus to use the word doggie actively means he has an idea of a class of dogginess. Late in his first year a baby will readily turn to things which have often been mentioned. This shows that the repeated concomitant use of a word in proximity to a thing has led the child to associate the sound with the thing. The sound has gained meaning; passive understanding is taking place.

It is generally agreed that talking, whether parroting or meaningful language, develops out of babbling. Babbling is sound–playing; it is a sort of fantasizing in that it is a mental activity in which there is neither a stable representation of anything nor does it obey formal rules. However, listening to a child suggests that there is much active and intelligent trial and error of mimicry and imitation in it. More exact research bears this out (Britton, 1985; de Villiers, 1977; Lee, 1979).

With playful trial and error, a child seems to approximate closer and closer to voicing recognizable words. When parents think they recognize a word they are usually delighted, hugging and kissing their child for his brilliance. The child for his part is then usually delighted with himself too, crowing with pleasure and repeating the word to get more of his parents' pleasure. Parents are perhaps not absurd to be so delighted; finding words is an incredible feat of perseverance and discrimination. After this a child usually races ahead trying out and finding many new words. Early speech usually involves comical mistakes or approximations which are nonetheless delightful, for instance: bissica for biscuit, blabbi for blanket, Wowies for trousers, toon for spoon, pisgetti for spaghetti, even coddispeeper for compost heap. These approximations indicate the *creative* invention that goes into learning to talk. Invention comes first, testing their conformity to rules comes second. Perhaps adults find such delight in children's early language because it is so freshly creative with a freedom that they cannot hope to match, engrossed as they must be in the routines and rules of adult life.

By the time he is two years old our child will usually be linking words into simple grammatical sentences. The psychologist Skinner (1953) has propounded that this comes about by a process of *reinforcement*. That is, the simple repetition of phrases becomes implanted in the mind by the pleasure of sounding the same as grown–ups. However the linguist Chomsky (see Britton, 1985) says this model is naive. A child does not just copy adults, rather he actively, of his own initiative, uses language acquisition devices, which are probably of innate origin. He *discovers* grammatical rules for himself. For instance, at first a child (usually a bit older than two) will imitate adults and use 'went' as the past tense of 'go'. But then, after discovering the general rule of past tenses, will use 'goed' instead. Only later will he revert to 'went'. Children seem to be largely independent of their parents' habits. For instance, adults will consistently say, 'I don't think it will rain today.' – not, of course, meaning that they are not thinking. Although children hear only this idiomatic form from their parents, they will use the more simple, logical structure which is, 'I think it will not rain today.' Only later will they take over the adult usage.

Chomsky has suggested that, no matter where a language comes from, there is a universal similarity of deep structures of grammar and there must, therefore, be an innate readiness to acquire these forms. How this might be, remains uncertain and there has been much argument about the question. Probably there are common bases to the structuring of thought generally, which rest on the innate propensities for mimicry as suggested in the last chapter. This thought structuring is then reflected in grammar.

With speech and grammar, the 'flattish' world of infantile

impressions is broadened and deepened immeasurably. By using words a child begins to be able to communicate precisely what his inner wishes and feelings are. To say 'drink' when he is thirsty is much more economical and peaceful than the crying and shouting he had to resort to before speech. By the time he is three years old he will speak coherently of his inner states. For instance, 'I feel sad, Jenny wouldn't play with me', or, 'We had a super picnic in the park'. Also, with language the child is brought into a clearer apprehension of other people's wishes and feelings. He will recognize his parent's wishes – 'Don't do that, it is very annoying', or, 'Mum feels tired now', or, 'Well done, that is a lovely sand pie'. Later still, the vast, moving and intricate web of other people's inner feelings will be opened up to him through everyday speech, novels, biographies, poetry and drama.

It is not only in the communication of feelings and wishes that language has such value. The realm of physical facts is widened immeasurably. By attaching known words to a new object, that object immediately gains meaning. For instance, walking down a strange road a mother points to one building and says, 'That is where they bake bread', and points to another, 'That is where the milk van has its garage'. With this, unknown buildings are brought into contact with familiar things and, in a flash, gain meaning for the child.

Language can, of course, in certain circumstances be condensed into signs and symbols to such a degree that they seem to be unrelated to ordinary words. This occurs in mathematics, where symbols condense what might take pages to write in ordinary language. It is perhaps a good exercise to try and imagine what our world was like before we learnt the meaning of words. With a little imagination we can get the impression of an inner and outer world where sound is only noise, where there are no sure signs to indicate hidden connections, and which is thus full of mysteries and terrors.

Child, Parent and Socialization

With the ability to walk and the mastery of language to express himself, a toddler must achieve a vastly enhanced experience of his separate self and his effectiveness. He seems to be working on and playing with trusting himself. These powers also give him a much wider appreciation of the world, its dangers and mysteries. It is a world of precipitous stairs, high balconies, hot irons, boiling pans, fires, curbs, grinding lorries, inexorable cars, mysterious ponds, weird neighbours with secret houses and of huge bosomy, booming people towering over you so that all you see clearly are their skirts, stocking tops, trousers and flies.

Thus, just as a child develops a sense of power, so paradoxically he must become even more aware, however dimly, of his helplessness. This sense of helplessness is an unavoidable consequence of growing intelligence. It cannot be said to be neurotic but rather an existential or *normal anxiety*.

What is naturally puzzling can be explained or further confused by parents. For instance: one mother might explain to a rather worried child faced by an eager dog, 'Yes he is rather big but I think he just wants to be friendly, just stroke him gently and see'. Whereas another mother might say about the same dog, 'Come away from that vicious creature'.

If his parents have been prone to distress him, then *mistrust* is likely to be a predominant mood. He may be able to discover a lot about the environment but *not identify* himself with other people in a relaxed, coherent way. Identification seems to involve first mimicry and then intentional imitation which is repeatedly played over and tested to become part of a child's self–possession. With mistrust, mimicry and ill co–ordinated imitations are likely to occur but, lacking relaxation with other people the structure of self–possessed identifications will tend to be broken up and discordant. For instance, a mother may confuse him by saying, 'Get out from under my feet. Go and play in the garden where there's lots to do.' The child then goes out into the mud and gets dirty, at which his mother says, 'Look at you, those nice, clean clothes all ruined. Come out of that filthy garden.' Here the child is put into a *double–bind* (Hayley, 1968) where doing something to please brings a parent's displeasure. This pattern has frequently been stressed by investigators of mental disturbance, it epitomizes how parents can confuse their children (Morrison, 1983; Olson, 1983; Parker, 1983; Reiss, 1981).

A child is not only faced with the mysteries of the physical world but also the contradictory vagaries of others' fantasies. For instance:

A child, crying and bleeding from being bitten by a dog, was smilingly told by its owner, 'He was just trying to be friendly.'

Some little children playing in a small park which was an old graveyard were terrified by an old man shouting, 'You little sinners the dead will rise and punish you.'

If a child is unable to split off, deny or ignore a vast amount of the fantastic verbiage of many adult communications he would probably be much more prone to overt confused distress than most children apparently are. But if crazy contradictions are insistently repeated by his parents he cannot escape their double–binds. Here are a couple more examples:

In a restaurant a mother says, 'Don't use a spoon for the peas, behave properly, sit up and use your fork.' The child sits up straight, picks up his fork, and with quivering hand lets the peas fall as he tries to get them up to his mouth. His mother then shouts, 'Oh, you hopeless boy.' The child breaks into tears which evoke the words, 'Stop that immediately, you are a disgrace and ruining everyone's dinner.'

A father, very proud of his pretty little daughter, takes her around with him in his car when visiting for his work. He is delighted by the attention both of them receive, but becomes irritable and scornful when she herself shows signs of being a flirtatious show–off.

Dreams, Nightmares and Phobias

The sorts of experience just mentioned are complex, contradictory and confusing. Each child's mind must have to work hard to make some coherence of them, however inadequate. It seems likely from recent research (Palambo, 1978; Wolman, 1979) that *dreaming* at night is one of the ways by which day time experiences are digested into coherently structured *memory systems* to be stored for future use. We know from examination of brain rhythms that dreaming, from before birth, in all mammals is a regular and frequent feature of sleep. How much lower animals assimilate experiences into long term memory is an open question. But humans certainly need to, and we would thus expect disturbed sleep to be associated with problems of assimilating experiences into memory. This is a fascinating and mysterious question. Little babies can have very disturbed panicky sleep which is probably associated with dreaming but they do not manifestly have *nightmares* which leave them awake in terror. These however do begin to appear at about eighteen months. They seem to arise only after a certain level of intellectual and emotional development has been reached. For instance, language is being developed apace at this time. Here, as we know one set of mental events (sounds) is being used to *stand for*, or signify, other mental events. Dramatization, as manifest in make–belief play, is another form of symbolization. Perhaps fully *dramatic* nightmares, which use symbolization, only occur after the child has actively begun to signify in a general way.

It is interesting to note that specific *phobias*, 'unjustifiable' fears of things, people and places seem to take root in children at about this time. These are in some ways rather like nightmares when awake, they involve the *condensation* of intense feelings into a *class* of things, a certain sort of person or animal say, or particular situations like dark places, very open spaces, or crowded areas.

Here again one sort of thing, a sort of animal say, seems to *stand for* some other set of emotional ideas in the child. Thus the coming of signification and symbolization seems to bring not only language but also dramatic nightmares and phobias.

Family and Cultural Context of Fears and Fantasies

Returning to more general considerations, it seems that not only does each child have his own idiosyncratic forms of activity, but also each family presents for him its own mixture of enjoyable impressions and useful information interwoven with confusing or stultifying, contradictory fantasies (Henry, 1971). Furthermore, not only do families differ in their competences and crazy fantasies, so also do cultures. Since each child is exposed to and has to adapt to a different environment from other children, he must structure his mental life differently from the child next door or in the next continent. He is being socialized in ways particular for him (Danziger, 1979; Dorr, 1978).

Vigilance, Rules, Conflict and Compliance

With mobility a child is into everything in order to discover and understand. A mother's presence is usually demanded but interference is not (Bowlby, 1969). However, we noted in the last chapter that such exploration not only presents dangers for the child (fires, cookers, electric plugs, stairs, roads, ponds, balconies), it also threatens a mother's precious possessions and habits. The looseness of primary maternal preoccupation has to give way to a more *vigilant preoccupation*. Fromm (1957) has distinguished unconditional love, which he associates with mothering, from conditional love associated with fathering. This is an interesting but perhaps slightly Victorian view; rather it would seem that a mother alone (quite apart from a father) is needed to be fundamentally unconditional in her loving during much of the first year of life, but more conditional in the second.

As a child is now very definitely active and intentional, the second year of his life is epitomized by commands, entreaties and threats: 'Don't pull the tablecloth', 'Come away from the fire', 'Please don't bring that mess in here', 'Where have you put that spoon?', 'Don't do that you'll hurt yourself'. This is a far cry from the Madonna and Child picture of the first year, yet it is just as central a part of mothering. It is very exhausting and essentially involves battles of wills. A toddler yearns to satisfy a multitude of striving for meaningful erotic, social and homeostatic enjoyments. And, when thwarted by his mother's words, gestures or stronger arms he is likely to be furious.

This fury is of a different quality from his paroxysms of distress in early months. Then, with very little sense of time and a rudimentary sense of self, frustration was likely to be felt as cataclysmic. Now, if it has been a steady and enjoyable first year, his more coherent and stable schemata of himself and his parents can be held in experience when feeling angry. Psychotic anxiety, where the sense of a separate self is lost or distorted, is thus not so likely to overwhelm him but be modified into a milder trepidation.

Even so, the switching of all–or–nothing reactions is still in the forefront. Any child when frustrated at this age may burst into an uncontrollable tantrum. The source of the frustration is often a mystery, it usually must be some misinterpretation or fantasy of what was going to happen. When this happens all parents can usually do is stand by rather helpless and embarrassed until their child is ready to let them comfort him.

Social conflict seems to be inevitable. If a mother inflexibly and coldly repeats forbidding gestures then her child may slip into a characteristic, bitter, self–assertive resentment. Alternatively, a facade of compliance may develop where vital spontaneity has died or is severely inhibited. Here something akin to a *false self* may have developed (Winnicott, 1958).

On the other hand, if a mother surrenders to a child's every whim, he is likely to be left with a fantasy of himself as magically powerful or *omnipotent*. This is extremely frightening because within such a fantasy everything is in danger of crumbling before him. It is a common clinical observation that children left quite unrestricted do not become freer but lapse into obsessive anxious doubt and depression.

The solution to all these conflicting claims seems to lie in a mother exercising sympathetic understanding. With this she recognizes, with pleasure, her child's inner vitalty and at the same time notices, expresses, and enforces if needs be, her own wishes and the needs of others.

Toilet Training

There is no need to say that this is a much–discussed aspect of socialization both by parents and experts. Everyone has opinions about it, but there seems to have been very little systematic study and observation of the subject (see Newson, 1963; J. Klein, 1965).

We know that at birth the nervous system is neither structurally developed nor functionally organized to cope with the complex organization that has to come into play to achieve sphincter control. Let us look at the sequence of thought that must take place in the child for control to take place.

He must first be aware that he *has* wetted or soiled himself. He must then become aware of the sensations *just before* he wets or soils himself. He must understand that other people want him to *communicate* this inner state of affairs to them. He must also inhibit the relaxation of his sphincters long enough for him to be brought to a pot. Lastly he must learn to relax his sphincters when over the pot. It is quite a complex sequence to learn.

Clearly a child does not necessarily learn the sequence in the order presented here. Many children, for instance, quickly grasp the last part – opening the sphincters over the pot – long before they master the first part of the sequence.

Early in his second year a child may simply not remember what has happened even a few seconds after he has wet or soiled himself. Engrossed in some new interest he will go on as if nothing has happened.

By about eighteen months he is likely to be able to say that he *has* wet or soiled himself. A month or so later he is just about able to say he is *about to* do so. After this he can begin to *voluntarily* clench his anal or urethral sphincter muscles.

By about twenty months he can begin to control himself to hold on and *move* to a pot. Toilet training is well under way. Incidently control of soiling usually comes before being dry.

One suspects that there are as many methods of achieving the sequence as there are mothers in the world. Perhaps we could distinguish two main schools of thought. First, the 'start it early and repeat it often' school who advocate potting from the earliest months to get the child used to the situation. Some mothers of this persuasion emphasize routine, and will put their children on the pot for half an hour or more after meals to establish a routine and a habit. The other philosophy could be called the 'wait till he wants to' school. This emphasizes that a child will learn quickly and easily when the situation of potting means something to him so that he *himself wants*, for his own as well as other people's sakes, to control his bladder and bowels. Those who wait for this usually have to put up with more dirty nappies for eighteen months at least. But they are likely to find that learning control has an element of fun and pleasure for the child as well as the parent.

It seems that the 'start it early' school are relying on a baby's proneness to void himself at certain times, immediately after a feed for instance. There is no harm in catching the motion then. But parents are under an illusion if they think a child is really learning to control himself. Relapses are common also until a child has really discovered he wants to be clean and dry. The final stages for this won't really begin to be assured before at least eighteen months during the day and of course much later at night (Leach, 1974).

Some General Considerations About Anal and Urethral Impulses and Fantasy

Potty training is not usually an openly difficult affair whatever method is used. Yet anal and urethral fantasy is omnipresently active not only in psychotherapists' consulting rooms but in everyday life as well. For instance, pleasure in dirty jokes is pretty well universal in spontaneous children who naturally get very excited about lavatorial humour. Later, having learnt to talk about sex, these are transformed into sexual joking, but the underlying lavatorialness seems to remain.

Less fun than joking is the continuation of anal fantasy into the habitual *character structure* of many adults. For instance, there is strong evidence that *obsessional* features like meticulous cleanliness, meanness about possessions, self–righteousness and preoccupation with obedience to rules do all tend to form constellations together as characteristics of some individuals (Kline, 1972). There seems little in common to bring these together except an underlying, often unconscious, concern about control of contents which at a body level is excretory control.

If toilet training itself is not usually a desperate struggle, why then should fantasies about it continue so troublesomely to cloud many people's adult lives? The answer to this seems to lie somewhere in the following direction. First of all, because the anal and urethral zones of the body are highly sensitive, they easily arouse erotic sensations and these particularly stimulate exaggerated fantasies. In addition, these zones mediate the expulsion of contents from inside the body. We know, too, that the body and its contents, in fantasy, are some of the self's most necessary and private possessions. And in toilet training some of these apparently precious possessions have to be controlled and then surrendered to another person. This being so, toilet training would epitomize, and generate exaggerated fantasy about, the *surrender* of aspects of the self and its capabilities to others. This problem of the surrender of self to others is not a trivial business for anyone, and for some is catastrophic.

It seems that *conflicts, originating in many other areas easily become unconsciously focused onto excretory functions* because these, being so excitable, are, above others, readily represented in fantasy (Abraham, 1927; Freud, 1905; Segal, 1985). One of the characteristics of mental activity at a highly emotional, often unconscious, level is seen to be working here. Functions that have a *similarity* are experienced as the *same* or *identical*. We have spelt this out before. Feces, body contents and the self are all similar in that they are possessions. However at an emotional fantastic level they are felt not just as similar but identical. The self is precious and feces and other body contents, being identical to the self, are precious too. What is more, if feces are surrendered to others, so too, at this level of imagining, is the whole self. When this carries on into adult life the idea of the self can be treated as if it is feces. This will seem a very weird, slightly mad reasoning to those used to everyday logic. But if one is closely involved with intimate emotional problems it will be seen happening every day for many people.

Another facet of the frightening importance of feces is their being fantastically equated with *dead* things. After death, animals rot and smell very like feces. A child quickly picks this up. As we have just noted in the last paragraph, things that logically are alike are fantastically experienced as absolutely the same. So feces and dead things are the same, and, since everyone has feces inside them, then they have dead things inside. Thence, using the peculiar logic of the imagination, our selves can be full of deadness, even ourselves can be dead! Perhaps this sort of fantazied imagination is at the root of the traditional Hell, this is never up in the sky nor on the surface of the body of the earth even, but in its bowels, in the underworld, usually entered through a hole in the ground, where devils eternally torment transgressors.

Freud was of course the first person to explore systematically the vagaries of illogicality at these primitive emotional levels of thought (Freud, 1900, 1915) followed recently by Matte Blanco (1975). Our example of anality and the self is just one more instance of the strange classificatory logic used by the emotions.

Returning now to childhood. It is not simply the child's fantasy systems that are involved but his parent's as well. He grows up, throughout childhood, fighting with and submitting to his parents' fantasies. Toilet training is only one event in a long sometimes benign, often chronic, history of struggle. Here is an illustration, evidence but not proof, of this contention.

A boy was brought up by parents belonging to a religious sect which was strict, rule–ridden and unquestioning. His mother was kind but a convinced believer and also meticulous, very busy and matter of fact, did not play games and was rather joyless, certainly not erotic. She toilet trained her boy successfully by routine from an early age. He remembers being fond of his family and passionately wishing to conform. But an underlying element of dreariness in life was epitomized for him by hours in chapel which were appallingly painful because he had to keep still all the time. However, the rebel was still alive in him. He secretly discovered how to explore erotically with pet dogs, and also how to play more open, ordinary games in fun with other children. When he went to school the relaxation of other families disturbed him and he became overtly rebellious. He seemed to express this at times by frequently peeing on the carpet outside his mother's prized and very clean kitchen at night.

As he came to adolescence the revolt became more open by refusing to accept his parents' ideas and challenging their beliefs and habits. Neither side could budge and at home he lapsed into a generalized sulking mood. Rebellious and intelligently free thinking in his general philosophy as he was, his parents' ways had nevertheless entered into him. He found himself in obsessive dilemmas about petty things that wasted hours of time, prissily meticulous about other people's faults and hairsplitting in arguments. He tended to find himself mean and over–careful about money. As an outlet he took to obsessive orgies of gambling, priding himself, however, on the meticulous, carefully worked out systems which he used before staking his money at the tables.

This illustrates not only a child's long–term struggle with parental habits but also suggests the association, which has been generally recognized for a long time by psychoanalysts, between early childhood control, rebellious aggressiveness, and *obsessional* thinking and activity (Abraham, 1927; Fromm, 1973).

Many other desires seem also to focus around discipline, which, you will note, is always *against* some immediate impulses, with the aim of forming controlled habits. Passivity, for instance, the wish to punish and be punished, desires for torturing and being tortured and *perversity* generally, all seem to centre around excitements about control and the desire to lose control. It is a vast subject for which we have no more space here.

Shame and Scorn

A child of this age, beginning to be self–conscious, is easily upset by his failures which are inevitably many and various. In the presence of others he is often obviously bashful or coy. And when he fails to do things in front of adults or older children, it is evident that he feels *shame* which he shows by hiding his face, running away or getting cross. If laughed at over his failure, *scorned* or shown *contempt* he shows every sign of being bitterly *hurt*.

It is odd that we should use the word hurt where no physical damage is caused, but when scorned we do feel almost physically hurt. Perhaps something of the following happens in these events. We are hurt most by those we love most. In love there is perhaps always a fantasy at least of finding a near–physical unity with the loved one. When that loved person is contemptuous the sense of a loving bond, with all its tender physical feelings, is savagely cut; we are dismissed or disowned and our self–esteem broken. It is as painful, often more so, than a physical wound. You will notice that here again is an instance of the strange workings of emotional logic. The self and the body are experienced as identical, both are hurt.

The experience of shame can be dodged by using mechanisms of defence associated with the splitting already mentioned in the last chapter. These are *projection and denial* (Segal, 1973). After these are used, the end result can be that a person feeling shame can convert it to contempt.

Here are a couple of examples of projection and denial.

I myself feel rather ashamed of being very incompetent at foreign languages. However, I am very competent in my knowledge of the London underground. When a lost foreigner hesitatingly asks me the way, I find myself sometimes welling up with a superior, scornful feeling of, 'You fool, don't you know your way around', and 'You can't even speak English properly'. I think I hide this well enough and perhaps am often oversolicitous, for I particularly dislike scorn.

Georgie Porgie pudding and pie, kissed the girls and made them cry, when the boys came out to play Georgie Porgie ran away.

In both these instances a person feels inferior about himself. Myself at languages; Georgie Porgie was and must have felt a coward. But this shame can be denied in perceiving weakness in others for it relieves one of feeling weak or incompetent oneself. I myself feel superior in seeing a foreigner flounder; Georgie Porgie presumably got a similar pleasure in making the girls cry. When an attribute *originally experienced about oneself is emphasized about someone else projection is said to occur*. Violent projective processes probably underlie all bullying, shrewishness, scorn and contempt. As will be obvious these malign activities abound throughout the world, not only individually but also seem to be particularly rampant when shared in groups, by cultures, nations and races. Projection and denial are here like twins acting hand in hand.

A process in the opposite direction from projection is also common; this is termed *introjection*. Here an attribute is first experienced as belonging outside oneself, but is 'swallowed' so that it becomes a possession of one's own self. It is a common occurrence when a person feels distressed by a loved one. Blaming a loved one is often frightening, so a person will readily leap to the idea that it was really his own fault. Blame is introjected into the self. In a less painful way, it is also probably used by children a great deal when they themselves feel posh if their parents do or have something grand. Introjection is a very common activity; it is less malign than projection because it does not unfairly attack another person. It goes on daily when learning skills from a teacher or parent. It is a fantasy associated with imitation and hence with identification. But it should be distinguished from full identification. This involves not only imitating another person and introjectively feeling like him, but also repeating and testing out to steady mastery the actual skills involved.

Lastly, in the context of shame, scorn and contempt, the feeling of *envy* should be mentioned. This occurs when an individual is unable to *introject* and then identify with something he wants to do badly, but still perceives it as the possession of another. It happens when imitation, introjection and identification fail. It can, of course, be corrosive because of the desire to steal or destroy the envied thing, but is also a spur to learning. It is hard to say how often it is acutely present in little children. In older ones it is very prevalent indeed, and, as the younger children are actually surrounded by people with much more than they have, it is probably very active in them too.

Primitive Guilt

Very close to shame is the experience of guilt. Both involve actual or fantasized awareness of being observed by others. They differ, however, in that shame is a reaction to failure to meet up to an *aspiration* of one's own. Guilt, on the other hand, may involve this but is essentially the experience in fantasy or fact (conscious or unconscious) of one's *effect* of causing *damage* or *distress* to another. The essential point is that it is concerned with the individual's experience of his effect on others. Psychoanalysts have stressed for many years that *fantasies of effects* on others are often much more powerful in experience than are actual effects. Such fantasies are not precise knowledge and thus tend to infinity of feeling.

Many psychoanalysts have pointed out (M. Klein, 1932; Winnicott, 1965), that primitive manifestations of guilt can be detected in the behaviour of children under two years old. One can see gestures of fending off expected retaliation, running away and also attempts to kiss mother better or to offer her titbits of food to placate her after a misdeed.

The credit for first investigating the natural personal development of guilt, rather than assume it as given of human nature, must go to Freud (1915, 1917). Since then academic psychologists have largely shied off the subject with a few but growing number of exceptions. Piaget (1935), for instance, has investigated the development of moral concepts in young children. Also, particularly, Kohlberg (1969) has developed and refined Piaget's experimental methods, indicating subtleties in the later development of moral conscience which Freud and other psychoanalysts have not articulated (see also Dorr, 1978, 1983).

Psychoanalysts themselves have not been very interested in statistical, experimental studies of guilt and conscience. Rather they have been concerned with their presence, genesis and habitual structuring in individual minds. Melanie Klein (1932), for instance, was the first to point out the omnipresence of intense, guilt-ridden, primitive fantasies in the imaginings of very young children which were paralleled by adult patients. The following discussion is drawn largely from her work; though in modified form, for she tended to envisage primitive guilt almost solely as a process internal to a child. This differs slightly from my presentation which envisages guilt as a product of social processes, coming particularly from parents, interacting with mental ones internal to the child himself.

The toddler's primitive guilt is, as Piaget pointed out years ago, premoral. That is, he has not developed a stable, coherent mental structure, of conscience which he identifies or possesses as his own.

It seems that in the first instance, as would be expected, commands, pleas and threats about a child's actions are experienced as coming *predominantly* from *outside* the self. This is natural, for it is from outside a child that they do in fact often come. We observed the precursors of commands in the first year of life in a mother's 'yes' and 'no' non–verbal gestures. It is in these and in the infant's reactions to the gestures, I think, that we must look for the origins of guilt. We have noted how a mother's 'no' gestures and rebuffs evoke depressed distress in an infant. This depressed distress seems to have fantasies akin to a sense of *death* in them, so we would expect ideas of death to be closely associated with profound guilt and remorse. This is certainly heard in the remorseful utterances of suicidally melancholic people. It is also seen more widely in culturally shared fantasies like, 'The judgement of sin is death', and 'Hell is for the wicked after death'.

As time goes by we would expect the young child to develop *expectations of punishment*, often fantastic in quality, based upon his experiences of a mother's rebuffs. This can be observed in little children cowering or hiding in expectation when no punishment has in fact been meted out. Here it is plain that, although punishments are felt as coming from outside the self, the child is beginning to develop representative *schemata* or mental structures about punishment or revenge that have a lasting quality.

This observation is confirmed when we note a slightly older child talking to himself, saying such things as, 'Don't do that John it's dirty', using his mother's voice. From our point of view, the child is clearly internalizing his mother. But, from the child's point of view, expected punishment still seems to him to come from the outside, not from within his own conscience.

In general, up to this age a child predominantly (but not exclusively) experiences his self consciously as a *passive recipient* of influences and effects from the outside rather than as one affecting others. This is natural enough because he is really small and powerless, very much at the mercy of huge, almost infinite, mysterious adults. His anxiety is largely concerned with what others will do to him rather than what he does to others. In psychoanalytic language it is said that his anxieties tend to be *persecutory*. That is, he feels his self as small, weak, passive and innocent in the face of powerful, unpleasant outside forces or persecutors. It is also said that, because he tends to experience himself as predominantly *passive* in the face of powerful, possibly cataclysmic, forces, he is primarily in a *paranoid–schizoid position* with regard to his image of himself and the outside world. This is perhaps not a very happy term for it suggests serious pathology which is not intended. It is used to indicate a normal developmental stage and to draw attention to the nature of the anxiety involved in it. This assertion that a child settles first into a passive position may seem puzzling. For it has already also been suggested that he probably has fantasies of his omnipotent effects on others, and this seems to contradict the assertion. But, while a child can quickly deny his omnipotence, being really small and weak, it is not easy to deny consistently his passivity, so he settles into that idea of himself first.

But, also as has been noted already, a toddler grows up, walks and very evidently gains confidence in his own powers and cannot help noticing his *active effect* in pleasing and distressing others. With this he seems to begin to experience *concern* about the, often fantasized, damage he has done. For instance, he kisses or hugs his mother better, or makes other reparative gestures. These are only rudimentary at this stage, but the work of internalization is beginning to take effect so that a new form of schema is developing. Around two years old or so, it is possible to hear little children say, 'Mummy, John will be a good boy today.' Then

86

later, perhaps at about three years old, more internalized, 'I'll try to be good today.' Because this process is primarily concerned with the *effect of self on others*, and consequently with self–accusations and remorse, it is referred to as a *depressive position*. Note that this implies a similarity to but is not identical with pathological depression. It is a normal valuable mental activity. Klein's distinction between paranoid and depressive functioning is a fundamental one. We can see them everywhere, everyday.

It is of note that, although the philosopher G. H. Meade (1932) was no psychoanalyst, nor particularly interested in the development of guilt, both he and Klein had a similar point of view about the sequence from passive 'me' to active 'I' in the development of a child's awareness of himself and his world. It seems that, rudimentary though it may be, the achievement of stable 'depressive' feelings is of the most fundamental importance. Genuine sympathetic kindliness and then the formation of a moral conscience seem to orginate in them. More than this, the very sense of self in a setting of a real external world seems, in part, to be dependent upon them. This is because depressive feelings are concerned with the self's effects on the outside world. And without an awareness of the effect of self on others, one is being only half realistic. Thus *concern about effects on others is an essential component in being realistic* both morally and intellectually.

However, this concern can be enormously exaggerated especially in children. We have already noted this in our earlier discussion of fantasies of omnipotence. So this concern and guilt can be very painful and frightening. All sorts of tricks or defences may be used to avoid full conscious awareness of them. Thus guilt is perennially defended against, but without its experience we cannot be fully realistic. This seems to be a problem not only for children, but for older people too. As any person comes to a new situation he is faced with the question of his effects on others. Painful fantasies of inordinate shame and guilt may make him avoid awareness of this. But if he does not face these experiences and *test out*, at least subconsciously, his effects on others then he cannot be realistic in the new situation. This depressive experience has to be 'reworked' at every phase of life. Without this happening, a person is weakened and probably prone to breakdown. It is also to be noted that with the coming of these depressive experiences more wishes and intentions are felt as coming from *within* the self. The feeling of things *inside*, of an *inner world* is getting stronger and, with luck, richer.

Defences Against Primitive Guilt

It has just been restressed that little children often seem, albeit quietly, to experience guilt in fantastically exaggerated, all–or–nothing ways. For instance:

A two year old boy seeing a bombed house said, 'Michael [himself] didn't do it.'

The play of disturbed children often shows obsessive preoccupation with fantastic guilt. Many less disturbed people can look back into their childhoods and remember dim, omnipresent, helpless feelings of responsibility and unworthiness, especially if their parents were unhappy in themselves. Here is a memory of fantastic guilt.

I remember myself, as a rather older child challenging my old grandfather, of whom I was very fond, to a running race which to my dismay he won. A few months later he died and I privately worried a great deal that the effort of the race might have killed him.

With guilt being such a painful, all–or–nothing experience it is no wonder that it is vigorously defended against in the mind (Freud, A. 1937; Smith & Danielson, 1982). One of the most frequent mechanisms or tricks (and the most malignant socially and personally) is the use of a combination of our old friends *splitting*, then *projection* with *denial*. In the instance of the bombed house just mentioned we can see denial operative. First of all Michael must have thought 'Did Michael do it?', then denial, 'Michael didn't do it.' In this case, the denial matched reality and his nurse confirmed it, but this does not always happen. Another instance shows projection operative:

A four year old boy was severely spanked by his mother for some misdeed and she shouted the words, 'you disgusting little brute.' He spent the rest of the morning interrupting his little brother in his play, hitting him and screaming, 'You disgusting little brute, look at all the mud on you, look at your vest, look at your filthy hands.'

Here, apparently after experiencing guilt himself, he spent the rest of the morning actively projecting it into his brother while he presumably felt clean and pure by comparison. Here, badness, originally belonging to the self, has been split from it, denied as in the self and projected into his brother, so that the boy can feel his self to be pure and good.

The splitting, projection and denial, the syndrome of, 'It's not me, it's him', 'It's not our fault, it's theirs', is, particularly pervasive in little children but continues on throughout the adult world. In psychotic form it becomes paranoia, as we have noted. It is present in 'goodies' and 'baddies' games, and thus in all goody–baddy hero–villain films which appeal to millions. It can also be seen in wicked, witch–fairy godmother, beauty–beast, God–devil fairy stories. It is, more seriously, prevalent in the group behaviour of chronically acrimonious party politics, in racism, sexism, class wars, belligerent nationalism and religious intolerance. It is very prevalent in private, but often equally painful, ways in marital conflicts. At an individual and marital level it has been much studied and is well documented (Dicks, 1967). It seems a great pity that, although we are beginning to understand so much of what happens in people's minds, the leaders and participants in horror–ridden social conflicts often utterly ignore what they are doing. Perhaps violent splitting and projection are essential ingredients of 'evil' activity anywhere. Our growing knowledge of these processes seems still to be virtually powerless to call a halt to people when they desire easy thoughtlessness. This comes naturally when splitting and projection hold sway.

FURTHER READING

1 Britton, J. (1985), *Language and Learning* (Harmondsworth: Penguin). A full but popular text.
2 Erikson, E. H. (1963), *Childhood and Society* (London: Triad/ Paladin). A classic study of child development from the psychosocial

point of view. A work that was formative for much later work, including this book. A must for anyone deeply concerned with development.

3 Kagan, J. (19891), *The Second Year: the emergence of self awareness* (Cambridge Mass.: Harvard University Press).
4 Leach, P. (1974), *Babyhood* (Harmondsworth: Penguin). Particularly for her description of toilet training.
5 Roberts, M. and Tamburri, J. (1981), *Child Development 0–5* (London: Holmes McDougall). This has already been mentioned in previous chapters.
6 Turiel, E. (1977), *Moral Development* (London: Fontana). A short summary of work on the subject.
7 de Villiers, J. and P. (1977), *Early Language* (London: Fontana). Shorter than Britton but very clear.

CHAPTER 6

Two to Three Years Old

Passion and Zest in Movement, Feelings and Ideas

Toddlerhood was epitomized by the discovery of a rudimentary self–trust in movement of the whole body. With this established, our child comes into the pre–school years. Vitality, full–blooded feeling and fantasy are the hallmarks of this period. In discussing this time of life it would be wrong to separate out each year too strictly from the others. Many things that apply to a two–year–old can also be said of a four–year–old, and vice versa. Developments are taking place at a rapid speed, one child will develop in one direction but not in another until later, whereas with another child it will be the other way round. However, we shall stick to age differentiation and highlight some functions which are particularly important at each phase, while remembering that they apply to the whole pre–school period and have consequences in later life.

This chapter will be particularly concerned with make–believe or symbolic play. The next will focus upon the group play of children and also upon feelings about adults.

With the firm knowledge that automatic movement is possible, the two–year–old will be seeking out new and more refined movements to achieve. From toddling, he goes on to walking steadily but, not content at that, will want to run everywhere, practise climbing, jumping and skipping. The most immediate impression one gets of a group of pre–school children is one of continuous flowing movement, eager, excited running, bouncing up and down from dawn till dusk.

This sense of physical autonomy is perhaps reflected in a child's conception of himself. As a toddler he referred to himself by his first name, which had been given to him by others. Now, less passively, he uses the word 'I'.

A young child's vitality is seen also in his talk, play and ideas. His chatter is predominantly a stream of observations: 'There's a bus', 'Ooh, a train', 'Look, a pram', 'I saw a cow today'. Later, at three and four, these simple observations give way to complex stories of events, and to questions, 'Why?', 'How?' and 'What?'

In both movement and ideas, a child will throw himself into one activity and then switch, often without warning, to something quite different. This flow of imagination, shifting from one activity to another, is exhausting to an adult as well as a delight. It is tiring not only because of the physical movement involved, but also because the adult, to keep up with the child, must allow his own imagination to range freely at the same speed as that of the child. He must give up his well–tried modes of thought and action and allow free play to his own 'childish' fantasies. To be childlike, and at the same time remain a realistic, responsible adult, is not easy; it requires being mentally coherent at several levels at once. However, to many parents the pre–school years of their children are unforgettable. There is enough drama of joy, anguish, comedy and violence from one family to fill a hundred theatres.

The two–year–old's capacity to speak gives us the opportunity to understand his point of view with a precision that was not possible in the earlier years. We see that, while a two–year–old is usually zestfully interested in the world, it is a puzzling place which often frightens him. He still rushes back to his mother when it gets too much. Mother or a well trusted surrogate needs to be kept well in sight except perhaps at home, even then she will be frequently checked up on (Bowlby, 1969).

As a child goes about the place he often seems to be talking all the time. His speech may be about what he feels: 'I want a biscuit', 'Don't want to go to bed', 'You are horrid'. But most often it communicates an observation: 'We did go to the swings today', 'We did see an elephant', 'There's the moon', and so on. Although stated as observations, these require an answer and a child will become very angry if there is no reply. He is not only making an observation, but he is also asking for confirmation that his idea is correct. He is trying, as we have mentioned before, to build up coherent and meaningful representations of the host of impressions that impinge upon him. A lot of his observations sound silly, but are perfectly sensible from his rudimentary point of view. For instance, many children search behind a television set or look under a telephone to 'find the people'. Or when told, 'We are going to post a letter to Granny', he may burst into tears at the pillar–box because, 'Granny locked in there'. Again, on seeing a chimney smoking, he puzzles and says, 'Bonfire on the roof'. Such observations indicate the feats of learning which every child must accomplish (Fraiberg, 1959; Isaacs, 1930).

Learning from Others

The child is usually quite content to have his observation simply confirmed by his adult companion. But he also relies on the adult to explain and fill out his direct perceptions. Obviously it would be a very careless person who just said 'yes' to the observation about bonfires on the roof. A child's parents and siblings are continuously required to correct, fill in and expand the child's knowledge of the world. Thus a child when drinking says, 'Milk', and his mother may reply, 'Yes', and no more – in which case his concept of milk is confirmed but not expanded. On the other hand, she may say, 'Yes – do you remember the cows we saw yesterday? Well, milk comes from cows'. The child usually readily picks this up, and may speculate with quite vivid approximations or fantasies as to how milk comes from a cow. He says: 'Cut cow open and milk comes out', or, 'Milk in cow's tummy'. This would be refined, perhaps, when the mother and child again see a cow, and she points to the udder and says it is where the cow makes the milk, and that it is sucked out through the teats by a machine and then put in bottles or cartons.

We have just noted how very seriously a child of this age needs coherent and amplifying replies from his parents. A lazy mother leaves a child ill informed, puzzled and with only his own ideas to go on; intellectual development is then very likely to suffer. An overworked or depressed mother, burdened by external or internal preoccupations can unfortunately also leave a child unsatisfied, puzzled and even frightened of the world. Still more confusing can be the replies and amplifications of a parent prone to paranoid ideas; suspicion and mystification can then intrude upon explanations given. The social chemistry of information exchanges varies from one family to another. Remember also that my illustrations are mostly British and children from other cultures will often be exposed to quite different information and misinformation. They are then likely to form different competences.

Mother, Father, the Family, the Outside World and a Child's Ambivalence

The physical movement of a child at this age indicates his eagerness to get to know the world outside his family. We have noted his fear if separated from those he knows well for more than short periods (Bowlby, 1969, 1973). But this does not mean he is uninterested in and cannot learn from other people. Many people will testify that other members of the family – grand-parents, aunts, uncles, siblings and nannies can become most valued objects of affection.

At this point it is important to stress the question of a child's ambivalence. We have seen how a child of this age is passionate; he can switch instantaneously from ecstatic idolizing love to vehement dislike of one and the same person. He shifts from all good to all bad experiences. This is *splitting*, we have already noted that it is useful in saving the immature child from confusion. But it seems that, if equable enjoyment and tolerance of other people is to develop, then a child must develop a structuring of ideas in his mind whereby he can experience good and bad feelings about one person *at the same time*.

It is one of the major tasks of growth throughout childhood to coalesce together into coherent ideas the conflicting ambivalences of loves and hates. Without it wisdom, intellectual and inter-personal, cannot be attained. For instance, an adult who gushes in adoration one moment and then is full of loathing or contempt the next is a tiring pest to others and often has poor intellectual judgement as well; he is immature. Such immaturity can happen anywhere, it is tied to no one culture.

It seems that close family members, mothers in particular, play a vital part in helping or hindering this tolerance of ambivalence. If a mother is present to receive and respond to all forms of her child's mood, good and bad, then he can slowly create an image of a person, his mother, who is good, bad and in between. She is then experienced not so much as either wonderful or terrible but as lovably interesting. Forming this representation of a person as a whole, with a continuity between the multiplicity of feelings of good and bad, sadness and joy, rebuff and gladness, takes a long time of exploration and experience. In times of extreme distress it

94

may shatter: for some children it is never remotely achieved.

Psychoanalysts stress that the young child's splitting and ambivalence means that he relates only to *part–objects* at any one time, a loved one *or* hated one for instance, not both. The process towards integrating good and bad feelings together is a vital aspect of what is called *whole object* formation. It will usually be years, if ever, for this to be stably achieved.

The importance of this development argues strongly for a child having at least *one* person (most frequently his own mother) to *both* love and hate, with continuity, at this age. If loving and hating is chronically split so that he only hates one person and loves another say, then the child's tolerance is not so likely to develop. This argument, however, does not imply that a child ought to have only one person to love and hate. On the contrary, if there is only one person he has no one else to turn to for comfort in moments of intense hatred. This is often intolerably distressing and he will resort to some distorting defensive trick to find comfort for himself.

The easy presence of another person, a father perhaps, allows a child to run to him crying, for instance, 'She's horrible, horrible. I'm going to smash her into little bits'. The other person can hold and comfort the child in a way a hated mother could not at that moment.

Quite clearly the role played by the other person in helping to resolve a child's ambivalence depends on the position taken. If a father, for example, replies by gesture of word, 'Yes, she's horrid, I'm much nicer than she is', then the splitting is exacerbated. On the other hand, if he replies, 'You are to say no such thing about your dear mother', then the child is left all alone with his hatred. But if he says something to the effect of, 'Yes, she's horrid sometimes but lovely too', then a child is probably helped to resolve his splitting.

This instance is in terms of a father as the 'other person'. It could equally apply to anyone else to some extent, an elder sibling or grandparent, for instance. However, a father can be particularly well placed for this function because he, above others, is likely to know love and be annoyed by his wife as deeply as the child, yet also can be dispassionate. He can share feelings with his child with both conviction and fairness (Backett, 1982; Lamb, 1981; Parke, 1981).

This argument is laid out in the language of a British nuclear family. In more universal terms, the necessities for maturing out of primitive ambivalence into more 'whole' experiences of people simply seem to be: at least two, mutually intimate and tolerant grown–ups. These can be found in family patterns other than the stereotyped nuclear one. However, even with people who have grown up in large, extended family networks in other countries, the nuclear triangle or mother–father–child within the wider network very frequently remains of vital and long–lasting importance to the growing person.

Girls and Boys

Along with all the other things to explore, a child will, of course, have been investigating his own body from earlier days of life. Contours will have been examined and erotic sensations aroused in all parts of himself. The main agents for this *autoerotism* will be the hands. With toilet training the encumbrance of nappies will have been laid aside, and our child will be much freer to explore the contours and sensations of the genitals. Here will begin the rudiments of *genital masturbation*. The *shape* of the sensations will be different for boys and girls just as the shape of a penis is different from clitoris and vagina.

It is usually about two years old or a bit earlier that a child shows signs of awareness of sex–differences. At this same time a child is also likely to become interested in the different body contours and colours of other people of all ages, particularly in the differences between female and male. Girls will notice and ask questions about similarities with their mothers and differences from their fathers; boys vice versa. There will be the first open glimmerings of *feelings of affinity* with those of the same sex and difference from those of the opposite sex. In a child's play one will notice girls often beginning to model themselves actively on their mothers, and boys on their fathers. The great class–distinction between male and female has begun!

At this point it is appropriate to summarize findings on the psychological differences between the sexes. There is a great deal of research material on the subject collected in recent years (Archer, 1982; Hutt, 1975; Maccoby and Jacklin, 1975; Ousted,

1972). Very briefly we can say the following. Until about the seventh week of pregnancy, human foetuses are identical in form, except of course for chromosome pattern. This form is essentially *female*. At the seventh week the male hormone begins to be manufactured and its action generates the development of male characteristics. Female development is not the same process, it simply occurs in the absence of male hormone. The development of male characteristics entails the following sex differences after birth. Boys tend to be stronger muscled, have a higher vital capacity and metabolic rate than girls. They thus tend to be more physically aggressive (in the sense simply of using their muscles thrustfully). Girls, on the other hand, have more body fat; this together with lower metabolic rate tends to make them less muscularly aggressive. Perhaps partly because of their imbalance of X and Y chromosomes, males statistically tend to *extremes* more than females. More male foetuses abort. There are more male stillbirths and more males born defective. In intellectual achievements there tends to be an overweighting of males at both the dull and brilliant ends of the scale (though this is undoubtedly coloured by cultural factors). In childhood, girls tend to develop intellectually earlier than boys, but this evens out in the teens. Girls tend particularly to be higher in verbal abilities, whereas boys excel more at spacial–motor abilities (this again may be culturally coloured).

Apart from these and the primary sexual characteristics of the reproductive organs, there seems to be little evidence of real differences between males and females. Ideas of superiorities and inferiorities have been formed over many centuries, fantasized, often shared between people to become prejudices, which are sometimes of a brutality that is now only too well known.

However it is also cruel to deny the differences between boys and girls. They are physically different and metabolize differently. This must mean that they *generate fantasies* differently because these originate in body sensations as we have noted. As fantasies are different so too are interests and ideas more generally. Those who argue that girls and boys differing play preferences, for instance, are just a function of cultural indoctrination are simplistic, they ignore the body functions that give rise to play. They thus do serious, often very hurtful, injustice to children's vital spontaneity. What is wrong with sex differences anyhow? How boring life would be without them.

Now let us turn to more intellectual aspects of play and exploration.

Symbolization and Preconcepts

In the last chapter we noted the beginning of signification not only in language but also in dramatic *symbolization*. In signification generally one thing or event stands for another. Many forms of signification arise through cultural convention as well as verbalization. We also noted a quite different form of symbolization which seems to arise more spontaneously from within a person, here the symbol *resembles* the form of the thing symbolized. Such symbolizing, most evident in dreams, seems to arise from the depths. Its source seems to be unconscious.

Piaget has stressed that signification is *rooted in imitation* (Piaget 1951). You will remember how mother–infant synchrony was an early instance of imitation. It is certainly present in a baby from the earliest weeks in automatic form. Later in the first year and on into the second we noted this imitation becoming more actively intentional. For example Piaget noted a girl, on seeing a child jumping up and down and thus moving his play–pen, soon started jumping up and down herself, clearly with the same intention of moving the play–pen.

An imitation that is clearly a *symbolic wish fulfilment* is instanced by a child trying to re–open a box and at the same time opening his mouth very wide. At about eighteen months the first symbolic imitations emerge, for instance, rubbing hands together to represent washing, or jigging running to be like a horse galloping.

Dramatic symbolization or make–belief proper has begun. Let us stress again that here imitation is occuring; a *body–action* is actively made to signify another event. Remember also that verbalization also rests largely on a vaste background of imitations; but, as we noted in the last chapter, words rarely have the *same form* as the event or thing signified.

Another point, in the case of verbal language a child comes to conform with general usage as he gains in skill. This conformity has a lesser place in dramatic symbolism, here personal invention remains more idiosyncratic. Even so the meaning of the symbols can usually be mutually understood by people or children dramatically playing together.

With symbolism comes the full beginning of interest in the *similarity* (as opposed to sameness or identity) of things (Piaget, 1950, 1951). We have seen in earlier chapters how synchrony and imitation seem to give rise to primitive feelings of sympathy, of being the same as another person. We have also noted the baby registering sameness in the first steps of classifying. With symbolization, from the second year onward comes a gigantically freeing step for the intellect. This is that *something the same is recognized as common to instances that are themselves different.* What is more this *attribute* in common can be *actively and intentionally signified*, either verbally or in dramatic symbols.

General ideas have arrived but they are not yet by any means logically consistent. For instance a child may easily distinguish a set of boys from girls and know that boys have willies and girls have fannies, etc. But put the boys in girls dresses and he will say they are girls. Or again: take a ball of plasticine, roll it into a long sausage and a child will say there is more of it 'because it's longer', or there is less of it 'because it's thinner'. These two examples show quite different sorts of pre–conceptualization, one is qualitative the other quantitative. But they both indicate that, even though our children have achieved the beginnings of general ideas and the use of signification about them, they have little idea of logical consistency. This inconsistency involves incomplete comprehension of the *constant* attributes that run through things that are similar. A child will be seven or eight before he gets into the way of checking his inconsistencies systematically even in concrete situations. But a start has been made (Beadle, 1970; Overton, 1977; Piaget, 1972; Turner, 1975).

Dramatic Symbolization or Make–Believe Play

Brief reflection about any sort of play will suggest that it rests on an active but relaxed manipulation of materials (in adults these materials may be purely mental). During play, distress must be at a minimum. The pleasures of play can be erotic but are more frequently those of meaning acquisition. From the earliest days of life, the infant plays in sensori–motor ways; so do other animals.

The dramatic or make–believe play we have just introduced is not confined to humans. It has been noted, for instance, that monkeys make characteristic 'let's play' gestures before gambolling together. In the absence of these serious fighting can ensue. Bruner (1976) argues that some measure of discrimination of a separate–self is necessary for true make–believe, even in animals. In other words, in order to play, a person must be able to experience, 'I'm me and it's it and now I'll *pretend* that it's something else'. Or if two individuals play together they must both be able to agree, 'I'm me and you're you, now let's pretend we are doing something else'.

Bruner and many others before him (Millar, 1968) have stressed the importance of play as essential in acquiring serious skills. Although oddly ignored by multitudes of parents and educators in the past, this is so well recognized today that further argument by me is unnecessary. Let us return to the young child (Cass, 1977; Garvey, 1977; Smith, 1984).

We have already stressed how, towards the end of the second year, the toddler will begin to state explicitly that such–and–such is something else, and then play with it *as if* it were the real thing. For instance, he will pick up a stick and gleefully say 'saw', and start a sawing movement with it. In the third year (when he is two, that is) he will probably multiply his instances of make–believe. Here are just a few. A stick can be a saw, screwdriver, fishing–rod, sword, gun, crane, hosepipe, wireless aerial, telescope, kerb to a road, aeroplane wing, or knife. A grocer's box can be a cot, bath, seat, car, train, aeroplane, boat, pig–sty, oven, table, cage or house.

Make–believe takes place in two directions. First, the stick or box is made to stand for something imagined. The visually recognizable similarity between a toy and the thing for which it stands makes this possible. It is here, of course, that the child is using his rudimentary capacity to generalize. What is more each symbol (*stick or box for instance*) here stands for, or signifies several *sets* or *classes* of things that *have similarity* between each other. But also a child pretends that *he* is somebody other than himself. Thus when he uses a stick to saw he is perhaps 'being Dad', or at least himself as a person who can use a saw. Again, in using the box as an oven, he is being 'Mum cooking'.

In make–believe there is an 'as if' suspension of reality. This is

100

naturally only partial. Children usually know very well that they are playing. We have already mentioned how both children and adults may on occasions doubt whether their fantasy is real or not. When this occurs, play ceases and distress takes its place.

There is something oddly paradoxical about play; it is an essential means of acquiring skills, in mastering reality, but in doing so a person stops bothering about reality to enjoy fantasy. The reality gains are a spin–off of the manufacture of fantasies.

Make–believe play is at least partly moved by fantasy, and this in turn is aroused by ever–present body feelings. One particular feeling about the body that underlies so much of a small child's life is his helplessness or incompetence. We noted this in the last chapter as a vague, not necessarily unpleasant, feeling of general or *existential anxiety*. With this, little children *play at being bigger or stronger*. Nearly always they play at being adults: big, strong, clever or grand. Their play is of lions, horses, cooks, nurses, dancers, mummies, soldiers, cowboys, doctors, tractors, aeroplanes, acrobats and cars. Sometimes they play at being little things like mice, but then they are usually clever enough to escape, or good enough to be preciously quiet. There is *wish–fulfilment* in play.

We see fantasy, probably arising from different parts of the body, in play. From the skeletal muscles there is play of diggers, cars and tractors and dancers. From the mouth there is lions, tigers and doll feeding. From the bowels and bladder there is messing and splashing, cleaning and tidying. From the penis, there is probably boys' enjoyment of thrusting, sticks, guns or spears. Fantasy from the penis is seen more evidently when a boy puts a broom between his legs saying, 'Look at my willy'. And from a girl's clitoris and vagina probably come fantasies moving her often to enjoy such things as dancing and also playing with the insides of things such as dolls houses as well as dolls themselves, flowers and frilly clothes.

But it is not simply body functions that are played over, it is the child *himself* in *relation to things*. When a child plays at being a digger or at being a mother with a doll, he can forget his smallness in relation to the world and for a time is its master or mistress.

This also happens in relation to people. What has been *passively* experienced can be played *actively*. For instance, having been nursed and fed by his mother, a child can play at actively nursing a doll. Having this in mind we can infer that active games of killing have been preceded by passive experiences of things going dead for the child. Play is, as it were a triumphant assertion of aliveness over deadness. Herein, perhaps, lies an argument against preventing little children playing battle games on the ground that they are too aggressive. In doing so adults may deprive them of the opportunity to play over and possibly find ways out of experiences of being made 'dead' or depressed. Battle games tend to be enthused about most by boys because of their interest in muscularity. Girls play of triumph over deadness is more likely to be about having babies.

Perhaps the morality of allowing some forms of play and not others should centre not on the content of the play but on the material of the playthings. We have, for instance, no reason to believe that dolls or little soldiers object to being knocked about. But when the plaything is a child it is another matter. When a spanked child goes and spanks a doll it is one thing. But when, as described in the last chapter, a spanked child goes off and sets about spanking another child without any 'let's play' agreement beforehand, it is quite another matter. Different rules have to be set for social as opposed to monologue play.

Let us now turn briefly to a general consideration. Play obviously does not cease with childhood; in the adult world creative play takes place with the minimum of physical objects, for instance a chess set, a writing pad, a drawing block or a musical instrument. Playing is embedded in all satisfying work, particularly in the design stages of a task. But the childhood need for physical objects in play often goes on into adult life without recognition. A child, being small, needs only a patch of mud to play in. Adults can readily enlarge this bit of mud to vast tracts of the earth in the urge to play their games – albeit unconscious and no doubt with rational or rationalized motives.

Adults and Children's Play

The arguments given above stress how children need time and space to play. The need to give physical *room in which to play* is now well recognized. More subtle are the limits set on play by adult attitudes. A child is prevented from playing over an activity not just by materials being unavailable but also by forbidding and rebuffing gestures spoken and unspoken. An, 'Ugh', or, 'Don't do that', prevents relaxed play as much as any wall. It seems only fair to a child that his adults should, except in emergencies, reflect, 'Why?', before they forbid a game. Otherwise a child is surrounded by incomprehensible fantasy–driven rules which are not realistic rules but whims.

But children *need* adults in play and not only as rule setters. Under about three years old, a child usually can only play solitarily or at most in parallel. But even at this age he does often seek to play with an adult. Parents on the floor and songs and stories at bedtime are usually remembered as times of bliss long afterwards. Adults who play with children are remembered with deep gratitude. This is not surprising because at such times a child can feel that an otherwise huge adult is a human with experiences like himself. The wholeness of the parent is enhanced. It is a time of sharing democracy. In particular, many of the frightening fantasies and puzzles which assail him in solitude can be transformed into fun when experienced with an adult who has similar fantasies but who is humorously unafraid of them.

In this, lie the principles of play–therapy and most other psychotherapies for adults as well as children. It seems more economical, as well as deeply satisfying, if parents can pre-empt such measures by having some fun themselves.

FURTHER READING

1 Archer, J. and Lloyd, B. (1982), *Sex and Gender* (Harmondsworth: Penguin). Covers more than the early years but valuable general reading.

2 Bryant, P. (1974), *Perception and Understanding in Young Children* (London: Methuen). A useful book which presents arguments against Piaget.

3 Fraiberg, S. H. (1959), *The Magic Years* (London: Methuen). A quarter of a century old but a classic elementary text about the psychodynamics of children.

4 Garvey, C. (1977), *Play* (London: Fontana). A useful elementary text.

5 Millar, S. (1968), *The Psychology of Play* (Harmondsworth: Penguin). More detailed but pleasant reading.

6 Phillips, J. L. (1975), *The Origins of Intellect* (Reading and San Francisco: W. H. Freeman). A simple introduction to Piaget.

7 Piaget, J. (1972), *The Child and Reality* (Harmondsworth: Penguin). A way to start reading Piaget himself.

CHAPTER 7

Three to Five Years Old

Playing with Ideas

As a toddler, even as a baby, a child will have been interested in other children without knowing how to get on with them. His play was *egocentric*. Around two years old, he begins to seek out the company of other children, but his games will still essentially be private. This has been called 'parallel play' or 'collective monologue' (Piaget, 1951). At this stage there is usually a lot of squabbling and fighting over toys, interspersed with quiet periods while children go separately about their own business.

At about the age of three, the first group games emerge. Here children share ideas, and help each other. It is social in the real sense, with *comradeship* and *friendliness*. This depends upon a child's growing capacity to feel himself as a coherent person in a world of other people, and this in turn is dependent not only upon his intellectual grasp, but also his capacity to love and identify with others.

In the intellectual sphere, bafflement about the world is still in the forefront of our child's mind. But now, having organized mental structures which take in more and more experience, he is beginning to be able to formulate puzzles in his mind and hence *ask questions*, which often go on from morning till night.

His skills broaden and become articulated. He will be able to ride a tricycle, probably actively keep a swing going. He begins to make recognizable drawings, scrawl 'pretend' writing and even form some recognizable letters. He will get absorbed in simple constructional toys.

Play becomes less focally concerned with learning how to manipulate objects. A child now enjoys things more for their ideational content. For instance, he tries drawing for what he can reproduce as well as for the pleasure of the feel of the pencil on paper and the sense of control that he is achieving.

Conceptual Development

Now, with his easy ability to talk, a child can make even more plain to us the nature of his puzzlement about the world. The questions asked by a three–year–old show how far he has got, and yet how uncertain he is in his conceptualization. Here are a few examples of three–year–olds' questions, taking representations and concepts of *space* first.

How big is a ship? Is it as big as the moon?

Can you hold a star in your hand? It isn't as big as an electric light is it?

Could I touch the sky if I stood on the roof?

Are clouds as big as a house?
(Cohen, 1982; Overton, 1977; Piaget, 1955; Turner, 1975).

He has a similar puzzlement over *time*. (For instance, 'We did see Father Christmas yesterday.' (said in late spring). It seems that 'yesterday' is often used for any time in the past, just as 'tomorrow' is used for the future in general. If you are a parent you have to be on guard not to promise a treat too far in advance. When you say 'next week' the child may expect it the next day and be bitterly disappointed. It is the same with unpleasant things, like being left or going to the dentist. Preparation too far in advance will only confuse a child. We can see again how full of infinities a child's ideas can be.

Number is similarly undifferentiated to begin with. The three–year–old can usually count up to two or more correctly if the objects are placed in front of him. Later if he can, say, count up to ten in speech he may nevertheless be unable to count this number of objects placed in front of him. He will forget which ones have already been counted and end up with a wild guess. There are great variations in this ability. Some children can enumerate quite young, but others are unable to count up to more than three by the age of five. The ideas of *constancy* in the external world are still shaky.

The statistical norms or average ages at which these many and various intellectual skills are developed have, of course, been studied by *intelligence testing* (Butcher, 1968).

At some time in the fourth year a child usually gets the idea that *written* words stand for speech and hence for things, actions or ideas. A few children can read and write by this age, but most content themselves with scrawling 'pretend' writing. They usually recognize a few letters and may be writing them in an unsteady way.

The child is also becoming more articulate in his representation of his own inner feelings and those of others. A few examples will show how intellectual and emotional developments are inter-woven:

'I don't like you, mummy, when you are angry.'

' "This Little Pig Went to Market", that is a sad rhyme.'

'David is cross because his mum smacked him.'

These are all about inner feelings, but their conceptualization and communication involve highly articulated intellectual pro-cesses. Think how complex these emotions are compared with the amorphous feelings of a baby a couple of years younger. Now the child is becoming coherent in *multiple sympathies*. On the one hand, this is dependent upon the child's ability to recognize that he himself is a coherent individual who has a physical place in time and space and has inner feelings as well. On the other hand, he is recognizing that other people also have their places in time and space together with their feelings. He has 'gone out', put himself in other people's shoes, identified with them and begun to see things *from their point of view*. Dramatic play or make believe must facilitate this. Many other intellectual functions must have played their part in the articulation of these feelings.

Optimally, intellectual and emotional development go hand in hand. However, it is plain that the two can become grossly split. Some clever adults and children are highly articulate in thinking about physical things yet remain sympathetically stupid. This can be seen from the difficulties they have in getting on with other people. The converse, however, is not usually true. It is most unlikely that an intellectually really backward person would also be emotionally and sympathetically intelligent. An uneducated person can of course easily be emotionally mature. But he must be intellectually developed to manage the subtleties involved in emotional relations.

Children Playing Together

Even quite young babies of less than a year show great interest in others of their age. They may even be carried away in imitation of them, for instance when they burst into tears at another baby's distress. They are also vulnerable to other people's moods, not only their parents' but also strangers' and particularly other children's.

Toddlers are more coherently interested in fellows of their own age. They often greet each other and want to be together. There seems to be a pleasure and relief in being with others who are the same size, whose faces are on a level with theirs, whose hands are as small as theirs, and who are puzzling about the same problems as themselves. But we have noted that their play is a monologue and egocentric. When paths cross there is often a fight because the two children want the same toy. Towards the age of three, however, children begin to want to play at the same game together. We begin to hear, 'Let's play at houses', 'You be ill and I'll be the doctor', 'I'll be a shoplady, you be a mummy'.

Just as solitary play is of paramount importance in intellectual development so mutual play is a keystone in social development (Isaacs, 1933). It seems to be vital in the growth of humane morality but is also subject to gross perversions: war, for instance, has with justice been called 'the great game'.

What mental functions are involved in this mutual play? The most obvious is language. As already mentioned, Bruner and others have highlighted the non—verbal language used by animals to communicate intention to play. The human child can, of course, do much more than this by using his speech not only to indicate intention to play but also in the content of the game drama itself. Incidentally, those who have had the opportunity to work with deaf or blind children will recognize the special difficulties that they have, although they and other children can be very ingenious in finding special ways to communicate.

But as well as using language, mutual play essentially requires a child to *sympathize* with others. He must be aware of them as human beings like himself whose needs must be recognized and accommodated to. There has to be a *sacrifice of egocentricity*. A child must have developed an ability to inhibit his own immediate impulse and whim for the sake of pleasure in communal enjoyment. One can hear this taking place in children's conversations: 'Oh all right, you get the water this time, I'll get it next.' You will note that the necessary inhibition here is facilitated by a complex co–ordination of ideas by the child. Without such intellectual achievements he can only remain impulsive.

Play, Conscience and Rules of Conduct

The ability to make the sacrifices of egocentricity at about three years old seems to be dependent upon the formation of a coherently structured, if still rudimentary, social conscience or *super ego* as it has been termed by psychoanalysts. We discussed the origin of this function in the last chapter. In order to play mutually, a child must be able to enjoy *sympathy* with and *concern* for others. We suggested that this is associated with depressive guilt. And furthermore, in inhibiting his egocentricity for the sake of the social game, he must check the expression of that rage and hatred which naturally arise when whims are frustrated. Thus social play depends on the working of conscience.

By thinking in a developmental framework, it becomes natural to infer that a particular child's internal structuring of conscience is dependent upon the long chain of earlier experiences with his parents: of pleasure, love, frustration and neglect which we have described in earlier chapters.

It should also be stressed that a parent's task does not end when a child moves off into mutual play with other children. Because children have established rudimentary consciences within themselves, they can now generate simple *rules of conduct* amongst themselves. But these only encompass short spans of time and can easily be swept away by the surge of a child's impulsive whim. An adult needs to be around to mediate sharing and to re–establish fair play. Children of three can be very aware of and sensitive to the fairness of sharing, but they have usually not yet developed the breadth of vision to re–establish it when it has been shattered.

The importance of the growth of mutual play for later life has already been stressed, but it can usefully be reiterated. In it can be sensed the germ of fairness and justice as well as social enjoyment amongst peers. It is a great step from dependency upon the autocracy of those early collossi, one's parents. It is the breeding ground of democracy.

It is of note that throughout much of history in both East and West the step to mutual play has been left to children themselves. Children seem to have tended to be left to find their own games while the adults got on with their work or pleasure. Development of fair. play has thus been left to young children themselves or their older siblings; a few rich families will have supplemented this by employing servants, nannies and maids to act as supervisors. But really serious consideration of these vital phases of play and development is only of recent origin. The pioneer playgroup work of Froebel, Montessori and Susan Isaacs is very recent history, and the formal recognition of playgroups and nursery schools is still incomplete in many countries (Fraiberg, 1959; Isaacs, 1933; Millar, 1968; Montessori & Carter, 1983).

Distress, Dreams and Play

So far we have only discussed the virtues of mutual play. But it can become dangerously destructive. To understand this we can start with our note in the last chapter that a fundamental motive for playing was distress and anxiety.

We have also thought earlier about *dreams*. These are now known from physiological evidence to be regular and very frequent in every night's sleep. There are certain clearly defined sequences of physiological response associated with dreams (these include, oddly enough, genital arousal such as erection of the penis in men). They probably play a role in establishing a record of our experiences in memory (Palombo, 1978; Wolman, 1979). It was probably Freud's greatest discovery to find that remembered dreams are coded or *symbolized dramatisations of distressing or otherwise significant experiences* that have not yet been assimilated (Freud, 1900).

It seems that daytime experiences are worked over, probably to go into our memory store, in sleep, if this is successful we wake refreshed. Sometimes a 'good' dream seems to have 'played' something over to a resolution and we slumber on with satisfaction. If the dreaming is unsuccessful in this way we remain disturbed and wake fretfully. Sometimes the distress produces nothing but a nightmare which leaves us puzzled and disturbed. It is incidently this characteristic of dreams to condense highly charged emotional situations into symbolic forms, albeit in a coded way, that makes them such useful tools for the psychoanalytic therapist.

Following Freud, it was later noted, particularly by Anna Freud (1928) and Melanie Klein (1932), that children's symbolic play bore similarities to dreaming. It often seemed to be a symbolic repetition of anxious situations. Further play seemed to have a settling effect. Children seem to need, are even compelled, to play certain games for themselves. A good game worked through leaves a child calm and satisfied, like a good night's sleep. A disrupted game leaves a child in distress. There is a very great deal that we do not yet understand about the mechanisms involved, but it does seem that *both dreaming and play can be means of resolving distress*. Furthermore, if a stress–provoked game remains uncompleted even further distress is created and the game tends to need to be repeated in an urgently compulsive way.

We can roughly distinguish two sources which might prevent the satisfactory completion of a game. The first lies prior to the game as it were, when the stress may be so intense for the child that his central control system or ego–functioning breaks down. In this case his mental activity is in such bizarre bits that coalescence into a game is impossible. This is not uncommon; a child can often be seen to be too anxious to play.

On the other hand, he may intrinsically be less shattered than this but prevented from playing by the limitation of external circumstances. These circumstances will usually involve his parents, for it is they who order his environment. They may explicitly forbid certain games, order him or carry him off to do something else, or they may simply provide a home climate where playing is devalued and poverty–stricken.

At all events, when playing is prevented for whatever reason, it seems that the frustration in itself creates further distress with its attendant hatred within the child. The need to play may then become compulsive with an underlying mood of tense fear and vengeful rage. *Persecutory feelings have swamped depressive concern.* Under these circumstances a child will not care about, or feel for, other things or people. He will want to use anything at whim; other children become *playthings rather than playmates.*

111

This is very common indeed, as was instanced in the description in the last chapter of a boy who cruelly smacked his little brother after being beaten by his mother. Here a game was played *without concern for the object* of the game, i.e. little brother. It is perhaps useful to term this *ruthless play* to distinguish it from genuinely *mutual play*. It is important to make this distinction for although both forms may have similar origins in a wish for mastery, they have very different consequences. Ruthless play has a paranoid quality. The mood behind it is full of frustrations which seem to be mocking and torturing, so that one wants to persecute others to even the score.

Naturally, just as with any symbolic play, this ruthless play does not end with childhood. The fantasies of such play are readily detectable in adult life even though they may be covered by polite or diplomatic smoke–screens. Ruthless play seems, for instance, to be rampant in all those engaged in imperialism, be it Roman, Islamic, Japanese, British, Russian or American; be it political or economic. Its most ghastly emergence was in Nazism. On a smaller scale, it can probably be detected in any activity where one or more people use others for their own ends without consent. Intimately it occurs within families, particularly in the stalemate tortures of unhappy marriages and also by parents who treat their children like dolls. The multiplicity of forms this play can take has been stressed by Berne (1964). His book *Games People Play* is light–hearted, often to the point of flippancy, but his message is a most serious one. The very achievement of comradeship in mutual play can be dangerous if ruthlessly used, for, two or more people together feel stronger than when separate. Their ruthlessness is thus all the more cruelly explosive.

It must be stressed that many personal ruthless games go on *without conscious awareness* of their theme, especially without awareness of personal hatred and its effect on others. Yet to the observer they are repeated with such consistency that chance or accidental occurrence must be ruled out. The theme of the game *is unconscious*; defences must thus be operating. Usually an array of splitting, projecting and denial manoeuvres can be recognized (such as when politicians adeptly lay blame on others for a belligerent war or economic catastrophe). But whatever the detailed manoeuvres may be blotting out or *forgetting* of the original stress, its anxieties and the theme of the ruthless game

112

seem to be omnipresent. This forced forgetting is termed *repression* in technical language (Freud, 1900, 1915).

Psychoanalysts have stressed that *unconscious* play is active in *psychoneurotic* symptoms. When suffering from these, the individual is not so seriously assailed by the disintegration of the differences between the self and external reality as in psychotic anxieties. These remain relatively intact, but our sufferer appears to be compulsively caught in highly charged ruthless games locked within his own mind. These may be grossly enacted with other people, as in the case of hysterical personalities. Or the dramatic themes may emerge less flamboyantly, not involving other people so much, in the distorted form of hysterical conversion symptoms (Cameron, 1963; Fenichel, 1946), phobias or obsessions. These of course deserve a great deal of attention in their own right but we can only point them out here.

The Oedipus Complex

Turning to less tortured, more ordinary activities let us briefly recapitulate the developments in the first three years. In the area of play we have seen a child begin with simple sensori–motor manipulations, and out of this, together with imitation, develops symbolization and the use of words. With this comes make–believe, which at first is a monologue linked at times into dual play with an adult. But during the second year our child becomes interested in and obviously feels affinity with other little children like himself. He now plays alongside or in parallel with others. Only towards the fourth year of life has a child developed enough to engage in truly social or mutual play.

At the same time we have recounted the growing articulation of a child's relationship with his parents. In earliest infancy there seems no distinction in the child's mind between what is himself and what is not. Only slowly in the first year does he sense his being a different entity from others. But even then his mother is largely conceived of as an extension of himself. Likewise, conversely, in a deeper and more longlasting way, he still feels himself to be a possession of his mother's. Nevertheless, differentiation is taking place, all being well. Especially from her 'yes' gestures and words of love and 'no's' of disapproval a child slowly has to accept that mother has a life independent of his own. With this comes the pain of loneliness as well as the freedom of solitude. In times spent together, however, it is usually his mother who administers his world; and even when he becomes interested in things beyond the home it is usually she who interprets and explains or confuses it for him. The childs's life is felt to depend on her; no wonder he is passionate in the extreme about her.

A father seems usually to be experienced in a similar yet also different manner. He has a different physical feel to him. It is he who goes out into the world and comes back bringing ideas and things that are different from mother. He is also usually bigger and physically stronger and this is sensed quite early, in the second or third year. Maybe there is a triumph for the child in the discovery of a mother's frequent dependence on her husband, something of a, 'Ha, ha, you have ruled me, now I've found someone who can rule you'. A child threatening his mother with, 'I'll tell my daddy of you and he'll smack you', is a common utterance. But then this can, of course, be easily contradicted in a child's mind if he hears his father being ordered about by his mother so that he is seen to be dependent on her.

I hope the previous chapters have given some idea of what giants, if not gods, parents are to a child. They are big, complex and mysterious because he is small, immature and full of trepidation about himself. Being at his parents mercy he will naturally often want to assert himself by manipulating and tyrannizing them. I think it is important to recognize that this *must be a given of the human child's condition* whatever our conventions of rearing are.

We have already argued that a two–parent nuclear family has many natural advantages for the rearing of young children. Later we will be investigating its weaknesses. To a child, even on the small scale of a contented nuclear family, parents are still mysterious giants. Add to this the frequent contradictoriness of most parents and the family must at best be puzzling.

It may be argued that this is an over–dramatized picture. A small, chuckling child shows no obvious evidence of his being baffled by his parents as complex giants. But we cannot expect a child to say everything that rumbles through the back of his mind. Furthermore, what can be more enjoyable for a small person than to live in the safety of benign giants who love him. Let us not forget how even we adults can find comfort in the idea of a benign God or in adulating leaders and royalty. Comfort can also be found in the converse; by a child or adult emulating the gods, kings or queens in manipulating, controlling or tryannizing others.

We have already seen how *bodily* experiences, tension of muscles, changes in blood supply, ventilation of lungs and so on are central to feelings; about parents as much as about anything else. It has also been noted that *erotic* pleasures are important in feelings. The very word 'feeling' has the double meaning, either personal experience of emotion, or touching sensations, which are of course intrinsic to eroticism. Erotic pleasure can work wonders with feelings, it can, as we saw in infancy, on occasion sooth anxiety, on others it can increase it by over excitement.

Having prepared the background we now come to the controversial conception of the Oedipus Complex introduced by Freud (1913, 1915b). This is centrally associated, in both boys and girls, with *genital* erotic excitement, pleasure and anxiety. I am going to give a simple outline formulation of this complex. It differs slightly from Freud, which I will briefly discuss later. Most present day psychoanalysts would, I think, essentially give a picture of the complex that is like mine. One or two aspects introduced by me, such as the role played by dreaming in its resolution, have not, as far as I know been stressed before. But by and large there is nothing very unusual about my formulation (Rycroft, 1968). Needless to say many non–analytic psychologists and psychiatrists still deride the whole idea of the Oedipus Complex. You must decide for yourself what you think.

We know that a child's stimulation of his body, particularly of the highly sensitive erotic zones, genitals, anus and mouth, arouses vivid fantasy. We have also seen how a boy or girl of about two years old begins to be aware of their sex or gender and of their similarity to the parent of the same sex.

Thus, being erotic, the genitals become particularly interesting. You will probably remember something of your own exquisite explorations at about this age. Genital play, masturbation, is pleasurable, comforting and evocative of fantasy but it is often forbidden or flustered about by worried parents. Most of all genital arousal, so easy to get going is *never* fully *climactically* satisfiable in childhood. Frustration with all its consequent persecutory anxieties is never far away (Lewis & Brooks, 1979).

There is one further function of the genitals which makes them very precious at this time. It is common knowledge among gymnasts and athletes that the 'centre of articulation' of the body in complex movements lies somewhere about the pubic area. Thus in our body image, this pubic, genital zone is felt to be at the axis of the organizing core of our athleticism. We know how important dance, gymnastics and athletics are to most little children and also how freedom in body movement is necessary for the growth of happy imagination. Another function associated with genitals has already been mentioned, this is that genital arousal consistently occurs at certain stages of the dream cycle in sleep. We have also indicated the probable importance of dreaming in assimilating memories and in resolving distress and anxiety.

Adding all these functions together, erotic pleasure, fantasy, gymnastic axis and dreaming, we can realise it is no wonder that the genitals are felt to be *the most precious organs* for a child. But apart from urination they have as yet no immediately apparent social use. What is more they are often surrounded by intense parental fantasy and taboos. For instance, sexual pleasure very easily can lead to 'mortal sin'. 'Don't fiddle with yourself or I'll get it cut off', is rarely openly said now, but expressions of dread, disapproval and embarrassment are not. Most parents never work out what to do about their child's masturbation.

To most adults the genitals at times seem to hold the secret of sanity and madness, 'Best anti–depressant there is, a good fuck'. At others they seem to be the root of our troubles.

How does this genital arousal and preciousness affect a child's feelings about his parents? It will be easier to answer this if we first distinguish the child's *inner sexual fantasies* about his parents from what parents and children *actually do* with each other. The former, the patterns of *private* fantasy, is referred to as the *Oedipus complex* (Freud, 1915b). The latter, the *social* triangle, is usually termed the Oedipal situation or *Oedipal triad*. The two are clearly related but not the same. We have already stressed how all parents relate to their children differently. Thus the Oedipal situation is different for all individuals. Since social experiences affect inner fantasy, every child's predominant and pressing fantasies will be different from others. Each individual has *his own privately patterned Oedipus complex* of feelings and ideas.

However, we can define certain common outlines based on the bodily structure of the two sexes. Consider first of all a little boy. As we have said, in *general non–sexual* ways, he both loves and hates his mother from all the particular experiences he has had with her. He also is beginning to feel himself as physically different from her (and she, of course, feels him to be likewise sexually different from her). Now, any ordinary, vigorous little child wants to love and hate with all the organs of his body. This includes his penis which sticks out. So he wants to love her with it, as well as with his other organs, and also, in hating her, wants to attack her with it. As it is a sticking–out thing he wants to touch her with it, stroke her, get inside her with it and have it cuddled. And also, in hate he wants, in *fantasy*, to beat her and get inside to attack her with it. These are urges or fantasies not often openly heard, except by a child when very excited or by children in therapy. Open expression of sexual fantasy is of course much less tabooed and more readily heard in an ordinary family today than it was a few years ago so what is being stressed here will not be entirely strange to many of you. But taboos still abound. The subject is still shrouded in mysteries and I cannot pretend that this description proves the universal truth about a little boy's sexual fantasy. However it is hoped that the question has been introduced in such a way as to suggest that at least such sexual fantasies could be important.

Let us now turn to a little boy's sexual ideas about his father. He, like mother, is loved and hated for all the child's past experiences. These too, have the urge to be expressed with all his body, penis included. However, he tends to see his father as being like himself; both are male, but his father is a giant compared to him. His father is bigger in every way including the size of the precious penis. So we have rivalry by son against his father. This is the celebrated Oedipal (or phallic) *rivalry*, named after the Greek tragic character King Oedipus who was fated to kill his father and marry his mother.

Lastly, a boy cannot help experiencing his parents being intimate with each other, both in love and anger. In particular they usually sleep together while he is *shut out* and *alone* or left with inferior beings like siblings. This proclaims to the boy that his mother is not his sole possession, sexually at least his mother is more intimate with his rivalled father than with him.

Being fraught with highly charged fantasy this situation of being shut out can re–arouse past non–sexual feelings of being *neglected, unwanted and scorned* with their attendant disastrous moods of depression, despair or hatred. Recognition of being shut out also wounds our little boy's pride. 'How can she do it to me, I am her most loved one.' It is one more blow to the egocentrism of the infant. It is because of its rearousal function that psycho-analysts stress the importance of this experience of being shut out when parents are together sexually. It is often termed the *primal scene*.

This experience cannot really be a catastrophe in itself. But it can be fantasized as disaster on the child's part in a recess of his mind. Nevertheless it has its recompenses; when he has had time, he can begin to feel relief that his parents are fond of each other. More than this the experience of *not being unique* is vital to a child's development of the sense of himself as ordinarily real in a real world. This is a painful necessity which most of us deny somewhere in our imaginations. How many of us, for instance, can easily imagine our parents making love together?

Some psychoanalysts have suggested that the sight of parents in bed together making love is a real catastrophic trauma for a child. This seems rather absurd, but what we can say is that a child is *very sensitive* about it and thus he should be treated with sensitivity. Most people can, at a conscious level, forgive and are glad of their parents making love together even though they find it difficult to imagine. On the other hand, parents who ignore their children's feelings and indulge themselves without thought in front of them are not forgiven. Brash, exhibitionistic sexuality disturbs a child greatly because he gets over–excited and cannot relieve himself. It is very often reported as a source of hatred and vengeful resentment for years afterwards. To counter this danger of over–excitement every stable culture, however open or naked the people are, has strict rules of *decorum*. We can easily see why it is right that it should be so.

Now consider the typical little girl with her different body structure. She, like a boy, loves and hates her parents non–sexually for her experiences with them. She, also like a boy, wants to love and hate both her parents with her whole body including her clitoris and vagina. (As one ordinary word for these organs does not exist, I will use the term genitals). These genitals do not stick out but are inside her. She must see her father as physically different from her even if she has little overt knowledge of his having a penis. And she, too, wants to love and hate him with all her body including her genitals. This means that as far as sexual love is concerned, she, in fantasy, wants him inside her genitals to love and hate him, either to caress or crush him in rage. But, of course, like the boy this is never orgasmically consummated, except in incest, so the urge continues in frustration until she finds a sexual partner in later life.

With regard to her mother, there is the same non–sexual love and hate arising from past experiences. However, physically and sexually she feels herself to be *like* her mother, but smaller than this magic giant. What is more when she sees her mother and father intimately together her smallness is compounded by realizing she loses her father to her mother. Hence arises the little girl's rivalry and envy of her mother.

Like the little boy, she seems quick to notice that her envied mother is smaller, physically less strong than, and often financially dependent on her father. This can accentuate one of two predominant fantasies. She can easily use this perception to deride her mother, as if to say, 'You think you're big but daddy is bigger than you and he specially loves me'. On the other hand, she can side with her mother, cool her envy with the idea, 'My poor mother and I are like each other, enslaved by this gross, giant father'. This is readily generalized to dislike of all men. Here we have the situation of *penis envy*.

This envy was considered by Freud (1905, 1915a & b) to be central to a girl's sexual development. He thought that boys' and girls' sexual fantasy was similar until about the age of three when the little girl realized she lacked a penis and hence envied all males because she herself was deprived of one. The formulation here is rather different from this. Most psychoanalysts now agree that Freud's theory of the girl's early sexual feelings is not quite correct. However, this does not mean that his concept of penis envy, or more generally of envy of masculinity is totally invalid. Much clinical evidence confirms that it is a consuming passion with some girls, but it is probably not so central in normal development as Freud thought.

Lastly the little girl has a similar sensitivity as a boy about her parents' intimacy and love making together while she is left alone. So the primal scene can be transposed from our description a few paragraphs back to apply to a girl as well as a boy.

Some Conclusions Concerning the Oedipus Complex

This ends the general formulation about the Oedipus complex. From it, I think there is no reason to assume that parental sexuality itself causes neurotic disturbances, except where the parents' behaviour has insensitively aroused the child's excitement in ways with which he cannot cope. This seems to happen when the parents have been brashly exhibitionistic or have intruded upon the child's genital privacy, as occurs in over–excitation, or at worst in incest or sexual abuse. Our description has instead pointed to a multitude of general experiences which arouse stress and produce inadequate methods of resolving it. However, the Oedipal situation *focuses* many old problems into one simple yet highly charged 'zone' of fantasy creation. The *genital core*, with its dreaming and playing functions, in the Oedipus complex make it central to sane resolution of conflicts. It thus becomes an axis for future development or disturbance. The genitals are the gathering ground of the child's early passionate life. Because of this, the common habit of laughing at the concept as old–fashioned seems absurd. To consider the Oedipus complex as the only experience of importance is blinkered: to ignore it is perverse nonsense. It is perhaps like any drama on the stage. To go to a play just for the dramatic climax of the last act makes a play essentially meaningless because the build up of tension has been ignored. But to miss the climax and resolution at the end makes the play equally pointless.

What is highlighted by the Oedipus complex is the apparent impossibility of resolving its conflicts and contradictions when seen from the impulsive child's point of view. For instance, a little girl wants to love her father with all her body including her genitals. But, quite apart from parental incest taboos, if she does actually experience him incestuously then the privacy of her genitals, with their sensitive use in essential dreaming, playing, resolution of distress, and probably memorizing, are invaded, and thus grossly disturbed. On the other hand, if incest does not occur, as is usually the case, then her genital yearnings remain irritably unsatisfied. It seems a dilemma. The same applies to a boy with regard to the satisfaction of the urges of his penis.

How can a resolution, or at least partial resolution, of these apparently impossible conflicts come about? We have just noted the importance for *sanity* of a child's genital privacy because of its close association with dreaming. We can then first of all see that the age–old incest taboo, usually upheld by parents, is not simply an outworn prudish custom. It protects a child's sanity and hence his emotional and intellectual development. Following on in this vein, we can also detect that the child himself, if he develops healthily, *wants* to surmount his erotic attachments to his parents because he feels instinctively that they tend to drive him mad. This is similar to the way we previously noted that a child reaches a stage when he himself wants to be clean and dry. To see how he comes to want to accomplish this second major feat of self–discipline, surmounting his erotic attachments to his parents, let us first look at his general social and intellectual development.

Inhibition of Eroticism

We have noted the pre–school child's all–or–nothing spontaneity, his eager, engrossed happiness. We have noted also his propensity to switch on and off quickly from one interest to another. This was termed normal splitting or proneness to dichotomize. This enabled him to be deeply absorbed and learn about minutiae very quickly, but it also meant that when persecuted he felt totally distressed even though it helped him to forget his troubles quickly too. The all or nothingness of his response also creates confusion for him, both intellectually and emotionally. For, as he grows older and learns more, new wider puzzles are created for him as smaller ones are solved. Splitting would tend to fracture his capacity to make overviews. We have already noted the healthy child's urge to solve puzzles. Piaget (1953) particularly has shown this in his observations. A child even more than an adult is always attempting to make sense of phenomena into wider and wider meaningful structures. In order to link up incomprehensible data he must integrate continuously, otherwise he remains puzzled and anxious. Piaget must be right in stressing that this urge to integrate meaningfully is fundamental, though it can very easily be disturbed.

We are thus presented with a paradox. The infant's proneness to split allows him to learn certain elements quickly, but by its nature it prevents integration of wider meanings. However if, after mastery of the elements, a child *waits* or stands back for a moment he finds that his mind can wander over a wider variety of phenomena and memories, he is then enabled to integrate them. The child has discovered that *inhibition* of impulse is *rewarded by the pleasure of greater understanding* or meaningfulness. Thus, in a normal child's growth from infancy onwards, there is a continuous dialectic between impulsive zest and a spur to self–inhibited appraisal. It can easily be helped or disturbed by parental pressures, fantasies and taboos.

We noted that this dialectic occurs in toilet training, but it probably occurs in many other developmental activities also. It seems to be very important in one particular instance. As a child grows up during the fourth and fifth year he usually finds pleasure in spending more and more time away from his mother in the company of other children. It has been noted that when a child is young these periods of being enjoyably away from his familiar mother are short, but when a child grows older he seems to become aware of the advantages of self–contained independence. He may be anxious about the lack of his mother, but when this anxiousness is inhibited or held in check, he triumphantly finds himself free to enjoy new experiences.

Here, through inhibition, we have the possibility of the resolution of conflicts that are consequences of impulsiveness and all–or–nothing splitting. Thus ambivalence tends to ease so that among other things the representations of his parents in the child become more meaningfully articulated, whole or rounded. Inhibition is frustrating and infuriating but it brings new freedom. In particular, the propensity for inhibition plays a vital part in a child's fateful steps out from his home into the wider, strange world, and especially to school. This is very important, so let us consider it further.

Latency, the Self–Containment of Oedipal Feelings and School Readiness

If you compare a group of children of, say, three or four years old with children of six, you will probably notice that the younger children are full of eager impulsive bounce but are easily given to tears and need to be hugged and kissed better. The older group, on the other hand, still show signs of bounce as well as tears but a subtle, more erect *dignity* has appeared in their posture. Their heads are, as it were, held higher, and they survey the world from a more *detached*, but still interested, point of view. They can view things from many different points of view. This *dignity, which involves inhibition*, not only manifests the child's wish to integrate wider experiences, but is also a sign of being able to be alone away from the comfort of his mother and her body. The finding of this dignity is a sign of readiness for school. It means that a child is able to stand the puzzlement and anxiety of being in a strange environment for as much as six hours or more at a time.

Going to school not only means being separated from home, it is also a journey into *new patterns of culture*. There is the culture of the classroom, itself presided over by the teacher. The child also meets and is called upon to relate to many other children who have come from the different cultures of their own families. Many of these will be like his own, while others probably come from different social climates and countries. Detached dignity seems very important in facing this *culture shock* more or less alone. As a child goes through the early school years one sees this detachment gaining strength. The period between about five years old and puberty, when detachment is growing, is often referred to as *latency* (Freud, 1915a & b).

Looking at this dignity rather more closely, we have already noted that the child stands more erect, surveying with detachment the world around him. The posture is very similar to the one a child will have seen his parents taking up when they were being protective, dispassionate and thoughtful. It is a mark of good parenthood. Psychoanalysts think that the child, in being dignified like his parents (assuming they have been), has not only found the stance for himself but is also unconsciously *identifying* with them. It is the fruit of long times of imitating them. They will now be felt rather less like mysterious giants and more like ordinary fallible and interesting whole human beings.

Recapitulating: this adoption of a parental posture, using identification, allows the child to be less swayed by switching or splitting from one impulse to another. It also allows him to need less immediate comfort when distressed. It helps him resolve, at least for the time being, his proneeness to the intense loves and hates which bind him to his parents. Inhibition and identification with his parents' dignity also helps him to identify with many other people, children and adults. He is thus gaining the rudiments of achieving the ability to see things from many points of view and reach an overview. This is the very essence of argument, debate and discussion: without which wisdom is impossible.

In summary, we noted in the previous section that the child's accumulated loves and hates about his parents tend to focus themselves into genital fantasies about them, and that these, especially, have a baffling and pressing quality. Now, with his detached dignity, helped by identification with his parents, a child has some means, albeit tenuous, of resolving these conflicts.

For instance, a girl in the throes of Oedipal feelings is swept by envy of her mother and yet loves and needs her too. When immature splitting prevails, this must seem an impossible conflict to solve. But with detachment the little girl can find that it is enjoyable to feel a bitter–sweet emotion of envy and love interwoven.

The Fateful Formation of Character in the Resolutions of the Oedipus Complex

The forms of partial resolution of infantile and Oedipal ties to parents are multitudinous (Cameron, 1963; Freud, 1915). Each child has his own pattern of loves and hates for his parents because of his personal experience of them. But also, the patterning of identifications, to find dignity out of those conflicts, is quite personal and idiosyncratic. Each child 'chooses' different aspects of his parents to both identify with and reject.

Each person's future is affected by events earlier in his life than four or five years old, as well as by his inheritance. What is more the course of his life will certainly be altered by later happenings.

But there seems to be something particularly poignant and fateful about the characteristic stance a child takes relative to his parents as he sets off on his solitary way into the classroom. One child will find his way easiest by proudly being his father's son aiming to follow in his footsteps. Another will 'settle' for a life that is 'morally better' than his father, or more potent. Another will naturally become kindly and nurturant like a good mother and take this on into his life with peers and adulthood.

The combinations of identification are near infinite, and later experiences will change them sometimes nearly out of recognition. But the threads of this early character formation seem to go on through life providing both strengths and weaknesses for life to come.

Let us take a few simple examples.

A self–assertive man married a woman who tended to do as she was told. They had two daughters. For reasons no one was conscious of, the father felt an affinity with one daughter, perhaps because she looked like him, and she naturally responded to his closeness to her. The other daughter missed this closeness. The mother, on the other hand, felt equally close to both girls. Now the first girl 'resolved' her Oedipal feelings by tending to dismiss her mother as dim, while being eagerly devoted to and modelling herself on her father. On the other hand, the girl less favoured by her father sided with her mother, accentuated her virtues, and seemed to reject her father as a male, chauvinistic martinet.

This is an oversimplification but it points towards how fateful the partial resolution of infantile attachments by identification can be. The form of the resolution must depend not only on the patterning of early experiences, and on the continuing stances of parents, but also upon the pressures put on a child by the environment outside his home. For instance, a boy prone to identify with his sensitive, artistic mother is likely to be hard put to establish a happy identification with her if he goes to a pugilistic, all–boys school.

Here are two more brief histories which show differing patterns of partial resolution.

The eldest of three brothers was the son of a kind but rigidly conforming mother and of a father who was a footballer. The boy himself was rather sickly as an infant, and his inexperienced mother was prone to react with horror because of her ideals of good health. Later, when his brothers were born, their mother became more confident and easy–going than she had been with him. The boy felt he was the sickly horror in the family and was frantic about it. Yet his sickliness had its compensations, his mother fussed over him all the more. He found his compromise. From the age of five or six he quite consistently chose the path of being the 'oddity' in the family. Everything that shocked his mother, and she was easily shocked, he found a pleasure in doing for he got more doting attention when being shocking even though he was being sadistic. His father on the other hand treated him with irritable dismissal and ill concealed contempt. Later, in his mid–school years, he loved to investigate everything that was 'different'. By ten or twelve he was absorbing himself in oriental mysticism. In his teens he naturally espoused any activity, philosophy or cause that was different from Western conforming standards. He was erudite upon Buddhism, Taoism, Zen, Shinto, acupuncture and macrobiotics. Athletically he could not be a beefy footballer like his father and brothers, but he found athleticism with a vengeance in karate. He loved to practice strikes which could kill.

This example, of course, does not exemplify positive identification with parents but rather a consistent life–style of rejecting both of them and all they stood for.

In her first years a little girl adored and was thrilled by her father. At work he was a carpenter, at home he was artistic and imaginative. However, he was irresponsible and tended to let the house go to ruin. The little girl's mother began to grumble about her husband and a chronic rift of bitterness grew up between them. By the age of six or so the little girl had at last become convinced, from her mother's arguments, that men in general were no good and her father in particular was a useless mess. Later, in adolescence, she entrenched her sense of contempt for him. In the early years of her adulthood she consorted only with women who likewise held men in contempt. Later, when forced to work with men, she tried very hard to hold to her old beliefs. It was only with great pain that she allowed her good feelings about them, which she had had as a very small girl, to find a place in her conception of them, herself and the world.

These were people from two countries in Europe. But the 'working out' of childhood family ties ranges much more widely and is a fateful question especially in cultures in flux. Here is a quote from China.

Mao Tse-Tung, speaking of his mother and his family, said: 'She pitied the poor and often gave them rice when they came to ask for it during famines. But she could not do so when my father was present. . . We had many quarrels in our home over this question.

'There were two "parties" in the family. One was my father the Ruling Power. The Opposition was made up of myself, my mother, my brother and sometimes even the laborer.' (From Snow, 1968).

This, of course, is not the case history of a crippling neurosis fit only for an analyst's couch, for Mao is humorously yet convincingly recounting the origins of his life's political passion and explicitly relating it back to his family as well as the social and economic conditions of his time. From our point of view, it stresses that personal Oedipal feelings may be fateful for an individual, even for a whole nation, but they cannot in themselves be reasonably considered as just neurotic or sick.

FURTHER READING

1 Freud, S. (1915), *Introductory Lectures on psychoanalysis* (Harmondsworth: Penguin and London: Hogarth). Freud's exposition about psychoanalysis specially for students. Long and detailed but fascinatingly vivid and fresh. It doesn't seem to have dated.
2 Turner, J. (1980), *Made for Life: Coping Competence and Cognition* (London: Methuen). If you have not read anything about cognitive development yet, now is the time to do so.

CHAPTER 8

Early School Days

Family and the Outside World

The law in Britain requires that a child goes to school at the age of five. This means that for six hours a day someone outside the family, a school and its teachers, takes statutory responsibility for him. Maybe for the first time in his life a child is formally under the orders, thus partially in the *possession of*, a stranger. Until now, unless he has been to nursery school, a child's awareness of the world, apart from TV, will have been filtered by his parents. He will also have attuned himself to their habits and morality and, since he knows no other, their standards will seem the only ones possible to him. Now he belongs to the school and teacher with different habits from his parents. The other children will also have unfamiliar habits and beliefs. Lastly, our child will be in a classroom to learn intellectual skills: he will not have submitted to this in a *formal setting* before (Fogelman 1983; Kelmer Pringle 1966; Newsom 1969).

Here are a few memories which distil some of the shocks of this new situation (Rosen, 1973; Tough, 1977).

> We were Welsh–speaking, but had to sing hymns in English. I know what they mean now, but can still remember the incomprehensible jumble of words from those school assemblies.

> I remember going to the medical room with a cut knee to have what I thought was 'flints' (lint) put on it.

> I jumped on a see–saw to show off to the Mother Superior of our convent who was standing near. The other end shot up and knocked her over. I was frantic because I thought I had killed a saint.

> I recall going to a strange school and standing petrified as the

other children bullied a boy with no hair. I can still see his reddish wig being thrown around and the boy crying with his hands covering his face.

I can remember my first day. I was all right but another child screamed when his mother left him. He must have been uncontrollable because he was put inside the wire guard around the fire, like a cage, and he screamed all the morning. I remember thinking, 'Is this what it is going to be like?'

The following memory from Nigeria highlights differences from these European instances.

There were few cars in our town and everyone wandered around and played in the streets. Most people knew each other. Yorubas are rather proud of being friendly, so you might spend the whole day out even if you were very tiny. Going to school was not seen as a great change and was eagerly looked forward to because it was very important to be educated. There weren't any birth certificates so you had an entrance exam. All those wanting to go to school were lined up and told to put their right hand over the tops of their heads. If you could touch the lobe of your left ear you could go to school, and jumped up and down with joy.

The shocks of going to school are of course well recognized and most teachers take a great deal of trouble with the early months. Yet however ready he may be and however kind the school, a child is often susceptible to being torn apart in efforts to conform to and placate the multiplicity of people and expectations which he encounters. Let us briefly examine the three main groups in this: the family, the teacher, and other children.

Early Home Development and School

The child carries the ways of home with him to school. These may have been happy enough in themselves but, to a young child's mind, they must seem to conflict with those of the school. This is particularly so when the school's national culture is different from the child's own. Here are two instances of dramatic change of environment, or *culture shock*, one involving national culture change the other not.

An Indian girl spent her early years in rural India. She naturally attuned herself to living in the open air and to the easy–going 'timelessness' of her particular village. When she came to England, school life seemed strange, cold and machine–like. Precision and timing seemed almost to attack her. She made many friends and was well liked. But her school days and classwork were strained and unhappy. Her teacher thought her a slow learner and she became seriously dispirited about her abilities.

An only child was much loved by the many elderly people in his family but most of his time was spent alone. He felt happy in this for, from a very early age, he populated his imagination with the people and things he loved dearly. He was reading at three and this increased the scope of his imaginative loves immeasurably. He normally never played with other children. By the time he went to school he was very well equipped to deal with both teachers and his lessons, but the presence of other children was quite outside his experience. They shook him out of his rich, solitary inner world of imagination. Because of his cleverness and his capacity to charm adults he became a teacher's pet, and thus hated by his schoolfellows who unmercifully bullied him. He never made friends with a child all through his primary-school years.

We can see that this boy was able to continue his amiability with the adults in his family across to good relations with his teachers. But his friendliness could not transfer to finding *comradeship* and *reciprocity* with his peers. This was a terrible, and probably unnecessary, price to pay for the teachers' regard and scholastic success.

Many parents seriously fail to recognize the importance of differences in habit between home and school. Parents who are strangers to a school's culture, as immigrants usually are, must naturally find it difficult to help their children with the change. But many parents without such handicaps display surprising personal lethargy and ignorance. For instance:

A mother was very worried about her son's health, and insisted

130

he wear long black stockings throughout the winter when the other boys all wore shorts. He was teased about this by the other boys, but dare not go against his mother. Later he found acceptance of a sort by becoming the clown of the class, but says he never forgave his mother.

Let us look at this situation of children together at school more closely.

Inside the Classroom and Out: the Society of Other Children and their Teachers

A young child at school is confronted with four different kinds of relationship. These are: the child himself and his teachers, the child and other children, other children and the teachers, lastly, other children together. The core aim in all these relationships is formal learning, or *scholarship*. Without this the essence of school is lost. However there are secondary functions in schooling which are vital to life. These are concerned with the social and emotional relationships of children together (Green, 1968).

The Society of Other Children

As children fundamentally wish to be enjoyed by their teacher as much as by their parents, there must naturally be an underlying competitiveness in any classroom. Poor performance relative to others usually leads to despair and listlessness in a beginner. Doing better than others can be equally unpleasant, as we have seen in one of the examples above, since a child places himself in danger of being envied and ostracized by the less successful. 'Swot', 'bookworm', 'teacher's pet' are often designed to distress the clever child. Such invective may be justified if a child becomes arrogant, but it can be meted out to even the most humble clever child.

Around this competition in the classroom, however, a child has the opportunity to find himself in the special society of others of his own age. This is often a crucial turning point for his later enjoyment of other people. Previously his companionship with

131

others will have been mainly under the close eye of a mother, or intimate adult, but in the school playground and afterwards in the streets or fields, he will be alone with others of his own age.

Children can be frightening because they may be strange, hostile and are less controlled and responsible than ordinary mature adults. They are fundamentally anxious and in fear of shame when they go to school. They will do all sorts of things to alleviate their *sense of inferiority*. On the positive side, they will be spurred to learn new skills, both in class and with other children. This creative competitiveness must be invaluable, but fear of inferiority may also bring less useful defensive manoeuvres such as bullying, teasing and scorn into play. For example, a child who is upset because he cannot read a passage in a book will feel better, for a time at least, if he sees someone else in greater difficulty. His self–esteem is enhanced by the comparison. This itself does not amount to bullying. For example, the private thought or even public comment, 'I can read better than Ann', might have scorn in it, but it is not teasing or bullying. But when a child actively repeats and repeats: 'Ann can't read, Ann can't read', until the little girl is in a paroxysm of humiliation, this is teasing, or verbal bullying. It clearly uses violent projection as a defence. It is clearly sadistic and can be as malignant or cruel as inflicting physical pain.

Togetherness in groups particularly lends itself to violent projection. Teasing and bullying is then transformed into *scapegoating*, when children gang up in mutual enhancement and obliterate their own anxieties in the discomfiture of another. For example:

An eight year old boy hears of a new gang that has been instituted at the other end of his street. 'Can I join you', he asks. 'Don't be stupid, it's against you, that's what it's for', he's told.

The fact that the members of a group get together in bullying means that they support each other in *obliterating* not only their fears of shame, powerlessness and inferiority but also their guilts. 'It's alright because we're all doing it.'

Needless to say this can become institutionalized; some teachers may even covertly encourage sadism by inciting children to scapegoat another 'for the sake of the school' or for some other cause, racialism for instance.

132

Apart from the dangers of violent projection, the company of other children is a rich ground for enhancing not only reciprocity but also whole networks of new relationships. A child can identify with his companions as fellow–sufferers on equal terms in the treadmill of school. Such *reciprocity* also gives a unique opportunity to develop further that *depressive aspect of conscience* which is based on sympathy for another human being (Klein, 1948). The roots of this may lie in early family experiences, but now it can be exercised in a wider society. This stresses, as Kohlberg (1969) has pointed out, that the development of conscience only begins in early childhood. Here perhaps democracy outside the family is born.

Teachers and Formal Learning

It is perhaps useful to discriminate two functions of a teacher. One is to introduce the ways of people outside the family to a baffled and anxious child, and the second is to preside over formal learning. The first task has been given much thought by infant school teachers in recent years, so that it is now often common for a child's introduction to formal learning to be slow and playful, with many intimate conversations between teacher and child. This is a great leap forward for freedom of thought compared to the dreary discipline of an old fashioned teacher drilling lessons into children.

However, just as parents often fail to recognize the anxiety a child may feel about the difference between home and school, so teachers can fail also. Children themselves are just not able to resolve contradictions by themselves. Both parents and teachers are called upon to help them with the transition. Teachers and other professionals are sometimes prone to attribute all blame for distress in a child to home circumstances, as it absolves them of feelings of shame and guilt. However, a child's anxiety must be a product of what he brings from home and the new environment. The latter may be just as inadequate for his needs as home. What is more let us not forget that personal consideration of individual children must only be peripheral for teachers most of the time. They are faced with a class of up to forty children and, as their main responsibility is to *foster formal learning*, many children must inevitably be left more or less alone with their own troubles.

Unlike home, where a child is usually left to choose his own games, school *learning is work*. Whatever means are used, the fundamental aim of schooling is to become *disciplined* in thought. Thus a teacher must inevitably be a disciplinarian, however subtle. She must expect to raise the antipathy that any inhibitor of self–willed fantasy evokes (Holt, 1982, 1983).

Incidently, as class learning is a new discipline so we would expect primitive fantasies and feelings to repeat themselves from earlier disciplinary times. Thus, recalling our chapter on the toddler, we might expect *anal* fantasies to come to the fore. We can often see these in children's private obsessional worries as well, more benignly perhaps, in the lavatorialness of junior school children's humour. It is probably for reasons of anal excitement also that the favourite old schoolmaster's ultimate mode of discipline was beating on the bottom, of all odd places.

Returning to more general considerations; old fashioned teachers perhaps saw themselves as needing to combat the unruliness of self–willed imagination and set out to destroy it early by the imposition of iron discipline and repetitive learning. More recently, teachers have come to see that this self–willed imagination lies at the core of every person's being and, if recognized, can act as a spur towards self–discipline. The teacher can thus become an aid to that part of the child which wishes to discipline himself in order to master a subject. However, there must be a part of any child's mind which loathes any form of constriction however benign. To this the teacher, however kind, must be antipathetic.

Old–fashioned discipline perhaps shattered a child's natural wish to discover for himself at his own pace. Modern methods do not commit this crime, but when the teacher is unwilling to face antipathy, then a child is left with inflated and unrealistic ideas unchecked. Free methods are obviously rewarding, but they also produce strains. A teacher must steer a difficult path between enjoying the pupil's ideas and at the same time checking wild fantasy.

Conceptual Development in the Early School Days

We are now truly into the region of education proper. But we have no hope of doing justice to its complexities, or to the emotional and intellectual problems that every teacher has to face. Even if I had the competence to tackle educational problems, which I definitely have not, we do not have the space to do more than brush their surface. However because it is a central argument of the book we will continue our discussion of Concept Formation (Boyle, 1969; Floyd, 1979; Piaget, 1972; Turner, 1975).

In the previous two chapters we have seen how a pre–school child was developing pre–concepts. With the coming of the ability to signify and symbolize intentionally he was beginning to be able to use general ideas. But they were full of inconsistencies and contradictions. For example a child of one or so going along the road may see a pillarbox and then a bit further on see another pillarbox. From his conversation it will become plain that he thinks it is the *same* box in both places. Not even pre–concepts have taken shape properly here.

Some time later he may again see two pillarboxes and by this time realizes that they are different but *similar*. This does not mean that he understands the logic of pillarboxes and how they relate to their super–ordinate class the Post Office. For example, we noted earlier a little child bothering that his Granny was shut inside one. By the time he is off to school of course a child will have disabused himself of such gross errors. He will however tend only to centre on *one* attribute or relationship at a time, so that logical *contradictions* abound in his ideas about the vast networks of relationships that comprise the real world.

Around three or four years onward and particularly when he goes to school we have seen how a child comes into a much wider network of other people than he has known before. We have stressed the importance of imitation from earliest infancy. We particularly saw its functioning in *dramatic–play* or *make-belief*. We also noted its more realistic functioning in *sympathy* with others and identification. This sympathy is a necessary step towards imagining the *self in another position*. From this other position one can of course, begin to imagine what things are like from that point of view. The wider social network of school allows for a great deal of exercise of this thinking in terms of *relativity*. A child thus has great new areas of reality in which to exercise his mind. In particular if he is lively, he will be imagining many new positions of himself relative to other things. *Self–object networks* are growing apace in the imagination.

In the last chapter we saw this with regard to a child imagining himself in the position of his parents. In this chapter we see this extended to other children and teachers. This is not only in the social–psychological sphere, but also in that purely intellectual world of things and their relationships. A child at school is now working out *complex networks of relationships* between *classes or sets of things* and individual things themselves.

Using his new facility to see things *from different points of view* he begins to be able to *conceptualize systematically*. This involves the child in exercising his capacity to be logical, albeit intuitively. The essence of this is to be aware of contradictions. Integration of different points of view, and hence breadth of vision, requires the *resolution of contradictions*; *logical* thought directs itself essentially at this.

Piaget suggests that after about seven years old or so a child is beginning to be concerned about consistent, non–contradictory, classification of things into *Hierarchies of Classes* and sub–classes. Asked to define a motor–car or horse, say, he will *try* and give the characteristics of each even though he may not get it quite right. He is concerned about logical consistency in a way that a younger child ignores (Piaget, 1950, 1953). He will, for instance be able to think about cars, say, as sub–classes of higher order classes, such as wheeled vehicles or means of transport.

This consistency is only partial, of course, and the child applies it selectively to one area after another as he goes through school. Furthermore he can only think in this way about classes of things that are more or less *directly manifest to the senses*. He cannot yet *actively* and intentionally *think* about *abstractions* in a non–contradictory way. Because of this Piaget calls this the stage of *concrete operational thinking*.

The purest simple form of consistent or logical thinking is, perhaps, arithmetic. This is based on the counting, or adding and taking away, of things. It is strictly logical and also *directly represents* touchable or countable things.

We can now summarize that both concrete operational thinking about physical things and socialized sympathy with different people involve conceiving of networks of different positions and relationships between them. Which comes first the social–logical or physical–logical is a very interesting question. My own impression is that they go hand in hand, the one enhances the other. It does not seem right to allot pride of place to either, at least in our present state of knowledge.

It is also important to note that even though conceptualization has begun to function logically, it simply adds to but never replaces more primitive pre–conceptual feeling and thinking; at least in the healthy individual. For instance any *inventiveness* or new idea always involves a *leap of imagination*. This must contain an element of dramatic play or make–believe. Such play must initially break the rules of non–contradiction. For instance, things that are really *different* must for a moment have been seen as *the same*. This is not logical but is nevertheless vital for thought. For example a teacher may say in an arithmetic lesson, 'Let us assume that this plasticine is dough and we are baking'. Here he is initiating a make–belief of playing at bakers shops. For the moment the children *are* bakers. However he then expects them to continue into a logical, non–contradictory mode of thinking within the make–belief.

The Development of Industriousness

As important as classroom work, is a child's gain in skill and knowledge in the company of other children. A moment's reflection upon boys' and girls' games, conversations and hobbies makes us realize that children probably glean as much information about the world from informal conversations with friends and family as they do in the classroom.

As he grows through the school years gaining, as we have seen, in the logic of concrete operations so his skills develop, and with them come awareness of *self–efficacy*. As this becomes more solid, the need for childish make–believe gives way to sheer pleasure in *industry*. We have just noted that make belief is still an essential first step, but it becomes subsumed under logically tested, non–contradictory realistic ideas (Erikson, 1963, 1968). Here is a memory which illustrates this transition:

I used to do a lot of tinkering and carpentry. When I was young I can remember pretending to be a carpenter or engineer when I was mucking about. It was fun doing it, but of course nothing ever really worked after I had finished. One day the handle of my mother's iron broke and in examining it, I realized I could shape a new one if I selected the right wood from the pile in the garage. I did this, screwed it on, and my mother could use the iron again. I didn't have to pretend any more.

The practice of pure explicit make–believe play perhaps falls away earlier in girls than with boys. It is my impression that it is boys in their early teens who are still absorbed in make–believe games and hobbies, such as model aircraft, cars, and so on. Girls of this age may have as many make–believe daydreams as boys, but on the whole their skills are often more oriented to reality. They certainly tend to be ahead in schoolwork.

This may in part be due to the fact that a girl is of the same sex as her mother and most of her infant and junior schoolteachers, whom she has seen at their tasks every day, will have been female. It is thus easy for the girl to imitate and identify with them, and hence to develop her skill and sense of efficacy. A boy, on the other hand, is of a different sex from his mother and probably most of his early teachers also. He tends to become antipathetic to identifying with them at quite an early age, yet usually has only a transient acquaintance with the tasks of men like his father. With less opportunity to learn he may be less sure of himself and hence more prone to pure, logic–free, wish fulfilling make–believe.

The School Child and His Parents

As a child gains more confidence to be alone he usually seeks out the company of like–minded children and turns away from his parents' conversation. It is often stressed that children no longer want their parents' company much during adolescence, but this progress towards separation must start much earlier. Children itch for the society of other children, which has a life of its own, with many secrets hidden from adults. Older children also like ganging up together to scorn and giggle at the silliness of their parents or adults. This naturally uses similar projective mechanisms to the scapegoating of other children mentioned earlier. During early school years this is usually transient towards adults so that a child can be laughing at his parents one minute and crying to them for help the next. It is probably not until adolescence that the persistent perception of faults in adults plays a central part in development.

The child's tendency towards separateness is only partial. A general consideration of these years makes us realize that parents' behaviour is still of overwhelming importance to him. The conscious focus of a child's mind may be directed away from his family, but fundamental decisions are mediated through it: he lives where they live, is fed and clothed according to their notions, and goes where they go at week–ends and holidays. Both economically and psychologically, he is incapable of independence. Children, by and large, are quite aware of this and accept the ways of their parents without open question or rebellion. This does not mean that the child acquiesces in all his thought and feelings. These will be kept in his inner world and may even be quite unconscious. At most, negative feelings will manifest themselves in nervousness or 'difficult behaviour'. His ego–functioning has not yet developed sufficiently for him to come out in open, self–determined rebellion. This must wait until adolescence.

FURTHER READING

1 Erikson, E. H. (1963), *Childhood and Society* (London: Triad/ Paladin). For the discussion of the development of 'Industry'.
2 Floyd, A. (ed.) (1979), *Cognitive Development in the School Years* (London: Croom Helm and Open University).
3 Green, L. (1968), *Parents and Teachers, Partners or Rivals* (London: Allen & Unwin).
4 Holt, J. (1982 and 1983), *How Children Fail* and *How Children Learn* (Harmondsworth: Penguin). Well–known personal accounts by a teacher based upon his observations of children he has known.

CHAPTER 9

Adolescence

Cast your minds back to your own teens. Maybe you remember: what you felt with your first period; the funny blissful buzz when masturbating for the first time; thinking about and comparing boys or girls you knew as you went off to sleep at night; your first argument with a parent on a matter of principle; the first time out late; feeling a girl's breasts or a boy's penis for the first time; your first smoke; your first political protest.

With memories like these in mind we can go straight into thinking towards what are the main psychodynamic changes in adolescence? These years are momentous for all of us. Yet all the social, emotional and even intellectual changes have roots in basic chemistry, in changes of *sexual physiology*. These biochemical transformations bring the capacity to *partner in reproduction* and to attain a body *size and strength* to work as an adult.

Sooner or later after puberty, a person is expected to take on adult *responsibility*, when this happens depends on the culture where he lives. Adolescence is recognized universally throughout the world and is often marked by instruction, initiation and ceremony. Though formal ceremonies are rarer nowadays, Western culture is no exception to initiation. For many children these years are ruled by school leaving certificates.

One of our earliest, known written inscriptions shows that concern about adolescence has been with us a long time. King Hattusili of the Hittites in about 1620 BC gives the following advice from a father to his son of thirteen. 'Keep thy father's word and stay away from wine in favour of bread and water while in youth, but when old age is with thee drink to satiety. And then thou may'st set aside thy father's word.'

In the West at the present time the pre–adolescent child *expects* to be cared for by, and be obedient to his parents in major issues. The post–adolescent adult in this culture and legal system cannot expect as a right to be looked after by his parents. Nor does he

140

expect, unlike King Hattusili, to be obedient to them.

Taking up adult responsibility is universal, but an adult's expectation of care from and obedience to parents varies very greatly indeed from culture to culture. Although the start of adolescence is fixed by our biological clocks at the beginning of puberty, it is drawn out and made flexible in its duration by both cultural pressures and individual needs. In our present society, for instance, there seems to be an expectation that adolescence ought to be completed by some time between twenty and twenty four years old. However, a person's actual assumption of responsibility often does not fit this norm. Adolescence may go on for many adult years in differing guises and is perhaps never completed before death. It is a process not a state; as such it is better, perhaps, to give it the form of a verb and refer to *adolescing*.

In these changes, many movements are taking place at once, there must be doubts about how things will turn out. Adolescence is a *process of transition* and hence a *crisis* for the young person and his family. It may be quiet and enjoyable, but I do not think it is possible to comprehend these years without having the sense of crisis in your mind. So let us start by defining this more generally.

The Concept of Life Crisis

This notion is a key concept in psychiatric thought when considering problems and breakdowns throughout the life cycle (Caplan, 1964; Erikson, 1968; Laufer, 1984; Mayerson, 1975), so let us briefly consider a definition of life crisis, which will set a framework for thought about adolescence and can also act as a model for other life crises.

A life crisis is a personal situation which arises when well–tried structures of adaptation and defence are no longer adequate to assimilate new demands, which may impinge either from within or from outside the individual. Loosening up with at least partial disintegration of thought and feeling then occurs. This is accompanied by anxiety and perplexity and often also impulsive action. Regression, with the re–emergence of gross primitive modes of fantasy, thought and feeling is likely.

A common term for these occurrences is 'going mad'. We have noted earlier that transient 'madness' can well be regarded as a necessary, healthy process. This because: *only when a person's previous structuring of ideas has broken into a fluid state is he then able to test out his widest array of responsiveness to the new situation. He can then discover new combinations of thought and action that are more satisfactory to his impulses, his conscience and the external world.* It could be put in another way by saying that *acts of finding new freedom* involve *transient madness.*

We have briefly defined the healthy state, let us now look at unhealthy ones. When an individual is faced by a new situation and is *unable to go into crisis*, then there is no breakdown but a limitation of personality or restriction of personal development. He remains *immature*.

Where a person goes into crisis but is *unable to re–emerge* then it can be said that he has broken down mentally. In such a breakdown, if the person remains fixed so that his self and internal fantasy are confused with external reality, then he is said to be in a *psychotic* state.

If, on the other hand, the person goes into crisis but fundamentally maintains differentiation between fantasy and reality, yet still *fixedly repeats his old ways* of relating self to others, while at the same time struggling yet not quite succeeding to adapt to the new situation, he is in conflict and is said to be in a *neurotic state*.

Where a person maintains fundamental differentiation between fantasy and external reality but achieves adaptation to the new situation through *obliteration of concern* for others, then his conscience or super–ego is *defective* and he is said to be *psychopathic*.

These definitions are oversimplified almost to the point of triteness, but they can be useful as starting points in your own thinking. They may seem heavy and overdramatic as a description of the apparently light–hearted transitions of life that many people enjoy through adolescence. It must be recognized that these are abstract and condensed analytic definitions, an actual crisis may last years, being slow, quiet and in the main very enjoyable with hardly noticeable acute distress or anxiety.

Let us now return to the young person. We can roughly give a sequence to adolescent changes, though people vary greatly in the age and extent of the emergence of each element, on the whole the order of the sequence remains the same. Later phases depend upon earlier ones (Blos, 1962; Erikson, 1963, 1968; Laufer, 1984; Matteson, 1975; Mayerson, 1975).

At about the time of puberty, or just before, some important developments in intellectual functioning seem to be coming to fruition. It is doubtful whether they have any necessary relationship to the physiology of puberty. But they do play an important part in the whole adolescent process.

Puberty itself is accompanied and followed by a physical *growth spurt*. This heralds changes in relation to ones *parents* at the same time as development of sexual activity and awareness. Changed relations to adults, together with intellectual developments, bring a new *moral*, often political, argumentiveness and awareness. Parallel to this, in the mid–teens also, often emerges individual aestheticism and with it new ideas of taste. Late teens and early twenties are marked by extending those activities and ideas which have by now become important to the self. Conversely a realistic young person finds he must at the same time narrow down immediate aspirations to fit in with money earning.

Let us look at these aspects of the process in more detail.

Intellectual Developments

You will remember how a young school child, no doubt under the discipline of his classroom, develops an array of *logically consistent concepts*. A fully consistent concept is one where no aspect of its meaning contradicts another. This checking for consistency was a great leap forward. Such concepts in the early years were closely tied to concrete data from the senses. However, many middle school children seem to be more and more interested in thinking *abstractly*. They are getting closer to *pure principles*; in science they will move, say, from ideas of constancy of objects to the principle of the conservation of energy thence to Einstein's relativity. In ethics they will move perhaps from ideas of fairness to consider the principles of natural justice.

Piaget has a clear and comprehensive theory about how this comes about (Piaget, 1950, 1972; Boden, 1979; Boyle, 1969). We do not have the space to describe it properly here. What is more this area of his ideas is particularly controversial. Academic cognitive psychologists on the whole take issue with his theory

143

(Hunt, 1982; Turner, 1975). Many educationalists, on the other hand, have found much positive inspiration in his ideas even when recognizing that his, like any theory, is limited (Shayer, 1981).

We will start with a brief indication of Piaget's thought, for it is the only one that can claim to approach the intellectual developments of our time in a deep and comprehensive way. Piaget and his school may be proved to be more or less wrong; but like all great movements in thought, our world is not quite the same again after them.

Around the age of eleven, or so Piaget detects crucial new developments in thought taking place. Firstly, a child begins to be able to think logically about the function of *several attributes* of the physical world or *variables* at once. Here is an example: a younger school child will have understood the concept of constancy of weight, how a weight stays the same even if you alter its shape and so on. He will also have recognized that weights can be added and subtracted. Even earlier in his life he will have become consistently acquainted with, say, another variable, distance, its constancy, addition and subtraction. But it will take time for him to *combine* operating with these two variables together in thought. Here he discovers that a weight and its distance from a fulcrum can be co–varied to produce the turning moment, or effect, of their combination. In other words, doubling the distance of a weight from its fulcrum can be balanced by doubling the counter weight at the same distance.

Many people will object that this is a task that many children under eleven can comprehend, it depends how much experience they have had with weighing scales. Piaget's age dating does seem too rigid, but the essence of his argument is that before you can combine variables in thought you must exercise them separately so that their consistency is understood.

With the combining of concepts (in this case weight and distance) into a superordinate one (turning moment) we have a concept that is more *abstract*, or at a higher level, than the two original ones. This brings us to the main feature of Piaget's description of the great intellectual leap. The young person is beginning to operate in thought at an abstract level. No longer do ideas have to be tied to concrete data, the visible, audible or touchable, to be dealt with logically. With the leap in combining concepts comes great interest in abstractions. And after this comes logical concern for the *non–contradiction of abstract concepts.* Piaget calls this the level of *formal operations.*

When this has occurred the logical mind is *freed from the concrete world.* Here the abstract speculator recognizes that, so long as clear specifications are made about a system of ideas, then it can be thought about logically in ways which bring valid conclusions. Such systematic networks of ideas need not conform to our external reality of space and time in order to be internally consistent. *Mathematics* for instance is logically rigorous but need not be tested against physical reality. A *scientific* idea on the other hand must be open to testing against the *data of physical reality.*

Mathematics is of course the most logically rigorous of our abstract disciplines. Algebra for instance is the abstract form of arithmetic where we are not concerned with visible or touchable things to count as in arithmetic. Rather, algebra is always interested in relationships between abstract entities. These have to be defined to begin with and then, if adequate, the logical consequences can be deduced. Thus it is perfectly possible, for a mathematician to conceive and think about a space with more than the three dimensions we are used to. 'N-dimensional' structures are a commonplace in mathematics. We adult or adolescent non–mathematicians can readily conceive that this is possible, but a young schoolchild would see no meaning in it at all.

Remember again that, what a young person is enjoying when he is being abstractly logical are consistent, non–contradictory, systems of ideas; here reasoning is by deductions from the definitions of the system. This is the enormous power of mathematical and any logical thinking. This logical consistency is essential in any science, but here, as we noticed a moment ago, testing against real data must also occur. All of us, scientists and mathematicians included, are, of course, only semi–consistent in our ideas at any one time. We have areas, depending on our skills, where we rigorously care about non–contraction and others where we do not.

Piaget (1950) suggests that the beginning of formal operational thinking marks the transition from childhood to adolescence. The adolescent, he points out, has an expanded time scale and thinks directly about the future. In being able to think and manipulate ideas unrelated to the immediate world about him, he can think about abstract possibilities. He can be an *idealist.* This has a paradoxical effect on the young person's relation to the real world. He is more realistic because he can well understand the relativity of differing viewpoints. But he is less realistic because being an idealist he is readily imbued with the *omnipotence* of thought.

Although Piaget suggests that this thinking and puberty come at about the same age it is a very open question whether one *depends* upon the other. Maybe they do, but many authorities argue against it, for instance the fact that precocious mathematicians master algebra long before their puberty argues against intellectual development depending on puberty. Conversely, no one suggests that precocious mathematicians are also advanced in their sexual development, puberty is obviously not caused by intellectual development.

Certainly the attainment of formal abstract thinking varies greatly from person to person. Possibly most people hardly attain it all. Some do in some areas and not in others and so on. Scholastic achievement, in the late teens, especially in the Sciences, does seem to have affinities with Piaget's operational thinking. For instance it is often possible to determine the conceptual level, in Piaget's terms, required for each physics or chemistry question in 'O' Level exams (Shayer, 1981).

A final note about intellectual achievement, it always seems to require rigorous discipline. For instance it is easy for us, when thinking of all the colourful emotional problems of adolescence, to forget that the years of the teens are for very many young people one long series of tense examination after another. This occurs particularly with so–called middle class children whose parents have high ideals of achievement. The strain of *internalized discipline* is enormous. Looking back and remembering the younger child's need to revolt against discipline we realize that it is no wonder that adolescents can go wild in anti–disciplinary orgies.

Remember also that, although disciplined logical thinking is essential to real *creativity*, it is not sufficient for it. Leaps of imaginative ideas, which need not be at all logical to begin with seem to be essential to any new departure, in science as anywhere else. It is likely that one has to be a 'bit mad', using the workings of our dream imagination, to see things in new ways. For instance, Einstein made one of his great leaps toward the theory of relativity when he was about sixteen and carried out one of his 'thought experiments'. This was to imagine himself as a particle of light travelling away from a planet at the speed of light, and then looking back at the planet thinking how it would appear. Identifying oneself with a light particle is hardly logical in the ordinary sense, but it was very fruitful.

From thinking let us turn to physiology.

Puberty

This, of course, refers to the maturation of *primary* and *secondary sexual characteristics*. In girls in Britain, on the average, breast buds and pubic hair appear at about eleven. Maturation is not complete until about twenty and the first menstrual loss usually occurs at about thirteen years old. The age of onset varies somewhat from country to country; in the USA it occurs a few months earlier than in Britain and in the tropics onset is probably rather later. At present the age of puberty is tending to get earlier; in 1900 it was nearly fifteen years old. This is at least partly due to less malnutrition and disease, conditions which are known to delay it. As we shall see in a moment there may also be psychological reasons for delay of menstruation.

In boys, pubic hair appears on the average at about twelve, and the penis and testes begin to develop at about thirteen but do not reach maturity until about two years later. The voice breaks at about the same time.

These changes are accompanied in both boys and girls by a spurt in skeletal growth. In girls there is widening of the hips, lengthening of body and change in body fat. In boys there is widening of the shoulders and increase in body length. In all these, of course, there are great individual variations (Archer and Lloyd 1982; Oustead, 1972; Smith, 1970).

These physical developments, hormonal as well as structural, introduce demands upon both boys and girls to change their mental image of their bodies, and hence themselves relative to others. Consider girls first.

Changes in Girls' Ideas About their Bodies
(Archer, 1982; May 1980; Mayerson, 1975; Schofield, 1965; Sharpe, 1910).

The appearance of breasts and the widening of hips are likely to be the first impacts of puberty. Depending on the girl herself and the cultural climate of family and neighbourhood in which she lives, these experiences can vary from fun in exhibiting her body to fearful shame and ungainliness. The greater sexual openess of the West at this time must surely be welcomed if only because it allows some simple natural enjoyments at a critical time of life.

The fundamental shift of puberty of course comes with menstruation. This means, not only 'the curse', with its worries about internal functioning, bleeding and mess but its consequences, 'I can have a baby'. This, together with the general development of her genitals, means that a girl is now likely to be swept with sexual feelings which can, of course, be delicious, disgusting or frightening depending on the girl. She cannot, from her past knowledge, have immediate means to deal with the feelings, however nice they may be in themselves, they are new, so some form of crisis is upon her.

For instance in extreme cases, it is quite well known that some girls cannot bear themselves as menstruating women with their fullness and bulging. They get into competitive slimming then, when a serious level of starvation has been reached, their periods stop. In a pathological way they have succeeded in returning to pre–pubertal asexuality. This is anorexia nervosa of course. Other girls, also worried about their attractiveness, comfort themselves with overeating and succeed in keeping boys away by being too fat.

A girl who is not too worried in these ways usually finds herself fleetingly but powerfully attracted to boys and men. Being unsure of herself she usually resorts, to begin with, to dismissing them. Conversations with best friends may then be full of delight in how silly or ridiculous various boys are, but the fascination still remains.

With this, probably more often in single–sex schools, come accentuations of lesbian–like crushes. These may be upon older girls or mistresses who seem to epitomize something *wonderful to be attained.* They are passionately *idealized.* On the other hand the object of devotion may be a girl who has a childlike, non–sexual, innocence of some sort. It seems here that a girl may be in love with what she feels she is losing for ever.

These crushes on females, which may only be fleeting or even virtually non–existent, are likely to be interspersed with bursts of passion for *part aspects* of men or other boys. A boy is distantly loved 'for his red hair', or 'his gentle face', or because of 'his firm voice'. This falling *in love with parts* is most publicly manifest, of course, in ecstasies about pop stars. In this we can hear the re–emergence of primitive modes of feeling and *all–or–nothing* emotions, they tend to infinity about simple aspects, parts or even bits of people. A similar sign of opening out into the *benign regression* in the normal crisis of adolescence can be detected, if less obviously, in a re–emergence of fleeting sexual feelings about parents. A father is quickly passionately loved then, equally violently, irritably scorned. A mother is bossily shouted at and then childishly made up to. Masturbation is likely to take on a new interest, we will discuss this later.

Boys' Ideas About their Bodies
(Archer, 1982; May, 1980; Schofield, 1965)

With boys too, the most marked personal feature of puberty is a genital excitability which they do not know what to do with. A boy is also usually swept by crushes on part aspects of girls, their breasts, hair, smile or posture. Not knowing what to do with himself, he usually takes to switching to boisterous mucking about with his pals and jeering at girls. Homosexual crushes burst out about part aspects of boys or men. Sexual feelings towards parents come up and, as with girls, are usually brushed aside with embarrassment. His body will probably have grown several inches in a year and he does not know quite what to do with it. He tends to be ungainly and irritably shamefaced.

Some Differences between Boys and Girls

For girls, menstruation comes regularly and inevitably from inside the body; it is not under the control of the will. The parallel

149

development for boys, ejaculation of semen, is controlled by the will and is, of course, played with to orgasmic pleasure in masturbation. It is possible that the boy's apparent psychological immaturity at this age is related to this difference. Menstruation very definitely cannot be the orgasmic fantasy game that masturbation is. Boys can go on playing with sexual ideas in fantasy. Girls do the same in masturbation but the clock of menstruation insists that this part of growing up is real. It is just not conducive to make–believe. Girls are thus perhaps rather more serious at this age than are boys.

Puberty used to be a time of intense sexual curiosity. My impression is that this is not so concentrated now. A child's curiosity is stimulated, and partly answered nowadays in the early school years of latency by television, press and conversation. So perhaps curiosity is not so urgent when puberty arrives.

Masturbation
(Blos, 1962, 1979; Laufer, 1984)

Close to the comfort of preoccupation with the image of the self, which we mentioned a few paragraphs back and will do again, lies autoerotic self–stimulation. This has already been noted as a source of pleasure, stability and comfort in infancy. Genital masturbation like any autoerotism seems to arise out *of frustration* which is close to moods of paranoid rage. The eroticism provides comfort and soothes the rage. The orgasm seems, if only momentarily, to wash all pain and rage away. However, as an activity it performs several functions as well as comfort. It helps the *discovery* of certain aspects *of one's own body* and brings its image closer to images of other bodies. It thus acts as a *step towards heterosexuality*. However, as we have just noted, it is also wish–fullfilling and thus a flight from the painful reality of physical frustration and loneliness.

So here we have another paradox. Masturbation helps to discriminate the body and maintains mental stability under stress. It is thus conducive to reality discrimination. But also, in wish–fulfilment, it is conducive to denial of reality.

It is most important to recognize that the patterning of an adolescent's or adult's masturbation fantasies *rest on a long history* of anxiety, frustration and attempts at self–gratification in childhood. In masturbation, the young person *draws fantasies from many zones of his body into the focus of the genitals*. And as the genital organs have a particularly satisfying quality it is very possible for insistent fantasies to become entrenched there. We noted this when discussing the oedipus complex in an earlier chapter.

Fantasizing in masturbation is usually about genital intercourse with someone of the opposite sex. However fantasies about other parts of the body may become the central pleasure. They may be for instance, about sucking, biting, beating, mutilating, urinating or defecating. The fantasist may be in an active or passive role. When *non–genital* parts of the body are *central* to the sexual excitement it is called *perverse*. The perverse pleasure may be solitary and masturbatory or it may be enacted with someone of the same or opposite sex. When entrenched into a central habit of life the perverse practice is definitively called a *perversion* (Rosen, 1979; Ruitenbeck, 1973).

Remember first that perverse excitements probably play essential functions in many couple's sexual foreplay, culminating in non-perverse genital union. Also do not forget that we are not making any moral judgement on perverse pleasures and acts in themselves. They may have seriously deleterious consequences, but so also can non-perverse sexuality if compulsively sadistic or destructive.

Let us go into this question of perversion a bit further. A provisional definition may be helpful. *A sexual act is perverse if, in male or female, the genital is aroused to orgasm by an object, human or otherwise, which is not the genital of a person of the opposite sex.*

By this definition oral sex by itself is perverse, so is anal intercourse with male or female. So also is mutual masturbation but solitary masturbation is not necessarily. Masturbation with a fetish object is, however, perverse of course. We will continue this discussion in the next chapter. All we are doing here is introducing the question. Our definition simply points up the centrality of two genitals being aroused together in non–perverse intercourse. Where this does not take place, in act or fantasy, then we are being perverse. This is not necessarily a bad thing. For instance the 'naughtiness' that there is in transient perversities may be very necessary to a healthy breadth of experience.

Denigration of Parents

We now come to a central psychological issue of adolescence especially as it is experienced in the West. Our discussion here focuses essentially on the internal changes in psychic structure, on the major shifts in patterns of feeling and ideas that occur in the years after puberty. It follows fairly closely the thinking of psychoanalytic writers, particularly Blos (1962) and Erikson (1963, 1968), but the attempt to integrate this psychoanalytic thinking with ideas about intellectual development, stemming mostly from Piaget, is largely my own and must be recognized as speculative.

There is also a great deal of literature based on social psychology (Fleming, 1967; Rutter, 1981; Schofield, 1965; Willmott, 1966). These non–analytic works describe and discuss how adolescents react to, and react with, the societies of adults they are about to join. This is one vital side of adolescence which must not be ignored. But I shall say little about it here for it is well covered elsewhere. Rather the focus for us will be upon the changing structure of feelings and thoughts within the individual as he 'adolesces'. Neither the psychoanalytical, intellectual, nor the social approach is complete in itself; one complements the others.

It was noted at the beginning of this chapter that the major psychological feature of 'adolescing' was the progressive *forsaking of a child's expectation of care from and obedience to his parents.* Implicit in the childhood attitude of expecting care is an underlying *idealization* of images of parents. This may not be overt, parents may even allow themselves to be humble slaves to their children, but, whatever the actual social situation, it seems that little children have to cling to a dream that adults have a sort of invulnerability. Only this will enable them to go on feeling that they are being safely looked after. A child may even be consciously disillusioned with his own parents, in which case he often creates an illusion for himself of invulnerable, good adults to look after him somewhere else. Without such an illusion children would be desolated and frightened. Such childish fantasy is enshrined in the wish–fulfilling fairy stories of prince charmings and fairy god-mothers. Only later is there appeal in myths of a less personal more abstract benign deity.

152

The sexual changes of puberty force upon a child the recognition that he is now reproductive; he is thus *close to* parenthood himself. In addition, his body is reaching the same size as his parents: girls are as tall as their mothers and many boys are physically stronger than their fathers. The penis or vagina is as big as the parents. These changes herald a mood of rather *manic grandiosity*; of 'I am as big as they are'. At this new level the young person feels *the same* as his parents, or rather he is the same as his parents were to him when he was tiny. For a moment at least he is *hugely wonderful*. It is termed manic because the big ideas have unrealistic elements. The young person has not been tested as to whether he can really function sexually, socially or at work. But he can be filled with fantasies that he can; sexual excitement enlarges these and masturbation provides a satisfaction of them. The manic mood is not only grandiose about the self it can very readily be also *denigratory* of his parents and other adults. Remember this grandiosity is just as pronounced in girls as boys. It is a culmination of the fantastic oedipal strivings of childhood.

This denigration is usually quite openly directed at the parents themselves, their beliefs, way of life and their discipline. It is infuriating and often most unpleasant. It seems, however, that, unpleasant as it is, the manic denigration *must* occur in some form if the child is to develop out of his childhood expectations of support from, and idealization of, parental images. Acceptance of parents' belief has, at least in part, to be *destroyed* internally in order for a young person to *work out his own values*.

For instance, as we have seen in intellectual activities, *argument* is dependent upon moving to *different positions* from which to view a question. Until a young person can be free to move into the position of demolishing his parents' particular view he is at least partially lame in argument. And furthermore any truly free thought is based on freedom to argue. So without being free to denigrate where appropriate there is no true original thinking. Incidently, this freedom to argue seems to have been referred to even in the Garden of Eden. Remember that Adam and Eve simply eat of the fruit of the tree of *knowledge of good and evil*. Like many exasperated parents, God threw them out of the garden. However our open minded conclusion now must be that this God was wrong to condemn Adam and Eve so rigidly!

The adolescent mood of denigration can also be justly called manic because, in his grandiosity, the young person *denies* his *own fears* of incompetence, helplessness, and loss of protective parents. In fact, it is well known that adolescents swing easily from arrogant grandiosity to helpless dependence. This was epitomized by one fourteen–year–old who screamed at her parents, 'Do something you idiots, it's your job to look after me.' It may validly be called manic and then depressive, but it is still the beginning of a process which is necessary and, if completed, is healthy.

Much of this denigration is often accompanied by *acting* deeds which assert grand independence and potency. This is evident in slamming the door and leaving home, sleeping around, impressing others by making a lot of noise, and dressing to be seen a mile away. New ideas, that seemingly do not come from the family, are being violently forced upon the parents. Grandiose assertion of power is also manifest in overt violence, in taking frightening risks like 'playing chicken' or in drug trips which terrify their more staid and conservative elders. There is a paradox in this acting, in so far as it is impulsive it shows that primitive splitting, paranoid and denial processes have the ascendance over more reflective, depressive and mature ones. On the other hand, without at least some definite action being taken, real testing of new powers can never take place. When acting continues without learning, depressive concern or testing then it must be deemed to be pathological.

Parents in Mid–Adolescence

In the face of insubordination and denigration which they have no real power to prevent, parents themselves are often thrown into bouts of despair. If parents humbly accept the adolescent's criticism then his grandiosity is confirmed and this usually makes him more frantic. For he knows dimly that his arrogance is not firmly based. He is also likely to be swept with depressive guilt about the hurtful power of his destructiveness which seems to reign supreme if his elders submit to him.

On the other hand, if a parent insists on continuing to order his son or daughter about as if they were children, then they still become madly angry and also contemptuous of their parents' delusion of authority. They know they are not little children and are readily disgusted by parental condescension. The home climate is felt as unrealistic. The young person is left with the alternatives of, either accepting his parents' words like a sheep and continuing in the false act of pretending still to be a child, or alienating himself as completely as possible from home.

A solution to this dilemma seems only to be found by parents going through a mini–adolescence for themselves. They perhaps achieve this by talking to their children as equally intelligent beings, which they usually are, while at the same time going over their own beliefs and conduct. At the same time they will probably find it helpful to address their own beliefs again arguing within and between themselves aiming for a new clarity and firmness for themselves. They will probably find their children are echoing many of their own questions that are still with them unresolved and put aside from many years ago. A child's adolescence brings the opportunity for his parents to be rejuvenated in mind if not in body.

Substitute Idealization

A void is left by the adolescent's destruction of parental images. This is usually filled, at least transitionally, by finding substitutes as models to be admired, copied and even adored. Young people may only have crushes on other boys and girls and pop stars, but they can become devoted admirers of older people, particularly of the same sex – uncles, aunts, teachers and young people's leaders. It can be more or less primitive usually with very little sexual involvement. When the older person is interested and interesting yet having a certain detachment it can be very precious. This is because, through their example, a young person can find the way to new positions and perspectives of action and thought outside the limits of his family.

We are nowadays, after adolescence much more the sum of our identifications with such people than just of our parents. Parents are our most primitive objects of identification, the adolescent cannot really expunge their effects. What he can do is destroy his own gross habits of internal childish expectations, and overvaluation and idealization of his parents. In this 'uncles' and 'aunts' can be invaluable stabilizers. If de-idealizing is achieved in the end, without the young person being alienated from his parents, then he can feel them to be nearer to friends, equal and reciprocal. His inner world is less under the sway of the adult giants of childhood and his images of his parents are further rounded, ordinary and whole. He is *free* to choose different paths by noting many identifications. But this takes a long time and is never, perhaps, fully completed.

Depression

A further consequence of attacks on parents' images are feelings of *emptiness* at the loss of old dreams with nothing certain yet to replace them. This creates depressive moods which particularly show themselves in the popularity of sad music about lost love. This theme of lost love undoubtedly echoes something deeply felt by young people. It may appear to be phoney because they do not seem to have had the time to lose any longstanding real loves. But they have: they have lost their naive childhood comfy love of their parents. Frequently a young person's sadness is quite consciously about old times past for ever. For instance, I remember myself returning at the age of fifteen to my old home two years after the family had left it, and finding myself uncontrollably weeping. No one had died, time only had passed.

Awareness of Death
(Anthony, 1971)

The young person's great sensitivity to *death* is probably also related to this sense of loss and mourning. He is now thinking more abstractly in terms of arrays of possibilities on a longer time base, as Piaget points out. With this abstract philosophical bent a young person can hardly escape conceiving more consciously of the end of personal time: death itself, not only for others but for himself as well. Feeling nearer to being an adult, a young person recognizes himself as closer to those who are old and about to die. These thoughts seem then to be reinforced by the 'internal deaths', such as loss of childhood, already experienced by the young person. We have suggested that these are to do with ending the age of innocent dependence on one's parents.

Reparation

Destructive attacks, even if they are only in fantasy, evoke *remorse*. Young people often feel they have *hurt* their parents badly; they do not usually really want to return them to their pristine state of idealized protectors, but they often make great efforts to heal their parents' apparent wounds. Depressive concern is essential in adolescence (Klein, 1948). One sixteen–year–old said, 'I have only two responsibilities, myself and my parents. I have a long future, they very little, I must look after them.' Such a sentiment is often spoken about in one form or another. Much adolescent rage is a fury at parents who will not recognize their child's attempts to wish them well.

The Self in Adolescence
(Blos, 1979; Gilligan, 1982; Laufer, 1984)

As a limbo of uncertainty prevails for a young person, a transitional security is gained by investing the image of the self with great importance; this is often referred to as *narcissism*. Being unsure of how and what else to love in safety, one's own self at least can steadily be attended to and loved. The 'face in the mirror' is a solace and helpmate. Even so, the image of self is not only pleasantly at hand for scrutiny and affection, it is also itself in a great state of flux. The adolescent is neither a child, nor the possession of his parents, nor an adult. He is a bundle of fantasies of possibilities. He is not yet someone with a set appearance or with an identity which has been tested in the society around him.

As the sense of self is so much at the centre of questions for the adolescent, it is no wonder that he makes it a focus of preoccupation. He is often in an opened–out, regressed, state where limits are blurred, thus infinity often pervades feelings. He is readily assailed by strange doubts: about his acceptability to others, about where he belongs, the presence of his body, the existence of his mind and the reality of the external world. These are worries of a psychotic nature; they are frequently transient and often go unnoticed by outsiders. To the youth himself they may be puzzling but carry no particular implications of great dread; they are part of everyday 'philosophical' thought. For some these psychotic processes get out of control and professional help is urgently needed.

Withdrawal and Mystic Experiences

Investment in the self and one's own body involves withdrawal, albeit transiently, from zestful love of external objects and people. This entails at least temporary ascendancies of *regression* into primitive feelings. These stem from when the self and outside world were less differentiated, when experiences were often more infinite, but also when all–or–nothing splitting held sway. Because of this tendency to split (especially into idealized all–good and all–bad), and to withdraw into self–preoccupation with its psychotic–like anxieties, it is often said that adolescence is a *normal schizoid* crisis of life (Jacques, 1970). This does not imply that adolescence is an illness but rather that, because *self–discrimination is so deep and vital* to the outcome of adolescence, the primitive schizoid forms of discrimination by gross splitting and crude withdrawal are necessary for its resolution. A dialectic between schizoid and depressive processes occurs repeatedly (Klein, 1948).

It also does not mean that the adolescent is perennially withdrawn. This is obviously not the case, rather, if all goes well, he moves back and forth, withdrawing to regain a sense of himself as unique and then opening out again to find new, enjoyable ways of relating to friends, the natural world and institutions. This coming in and out of withdrawal is seen also in the need to stay up all night and to sleep a lot.

In this movement between inward and outward there is often a sense of *gazing in two directions*. It has a poetic quality to it and in it may lie the key to much creative activity, especially in the arts. Articulate adolescents often boil over with interesting ideas. However they are usually not in the position to be uniquely devoted to one medium. They have not had, and maybe should not have, that opportunity, which depends on restriction, to spend long hours and months in refining *techniques* to communicate their visions in the economical form demanded by fully mature art. Good creativity is after all 'ten per cent inspiration and ninety per cent perspiration'. If the adolescent does narrow himself down he is impoverished in his explorations and will have little richness of experience from which to draw for his ideas.

Returning to the idea of gazing in two directions, one of these ways is to look outwards and be very open to sensations from the world around. The young person quite often can experience sudden mystic feelings, often with an infinite sense of unity with the universe, when things are seen very clearly in a flash of great beauty. One young person said:

> The things around me were of great beauty, they had no yesterday nor tomorrow, just now, without time, there was no urgency, they were just there very open and clear. I suppose it was how a dying person sees things.

Or, it might be added, like perhaps a new born child sees them. Psychedelic lighting and sounds could be attempts to stimulate these flashes of clear, pure vision; drugs certainly are. My impression is that neither method of induction has the pure ecstasy of the mystical sense of beauty experienced in the natural course of life, for both psychedelics and drugs have a forcing quality about them. And drugs are also poisons.

Turn now to the other direction of thought, inwards. Our adolescent is, albeit transiently, very aware of his inner happenings, of the delightful new harmony of his body but also its clashing un-gainliness. He is, perhaps more dimly, concerned with the function-ing of his mind, its new found powers of abstraction and the frightening craziness of unsought ideas that come up from nowhere.

It is perhaps in mystical, religious and aesthetic feelings that a sense of union occurs between outward and inwards. When one's inner harmony, for instance, melts into a sense of harmony about something in the outside world then we leap with joy at its beauty. When a tune, say, echoes or means something to us, we do not feel alone – the melody and its makers are at one with something within us. Likewise when we experience a synchrony, not just between oneself and a thing, but with all things, all creation, then we are in a mystical experience.

159

Somehow *abstract thinking* plays vital roles in deep aesthetic and mystical experiences. For instance participation in both activities seems to have been confined throughout the ages to people who enjoyed 'high' thought which always involves abstraction. Profoundly religious feelings also involve abstract thought particularly of a deep moral and ethical kind.

Why abstraction and these deep experiences should go together remains largely a mystery, to me at least. Perhaps one aspect is that it is through abstraction that we 'grasp' many individual events together in our mind. Our mind thus 'soars across the universe' and we are nearer, at least in imagination, to embracing the whole of it. We are then closer to being 'at one' with it.

It is only with a rich, 'abstraction ridden' and turbulent adolescence that such ideas are deeply meaningful.

Conscience

We are now in a position to draw some of the strands of this chapter together. It has been suggested that the changes of puberty provoked a somewhat grandiose tendency which involved denigration of parental images. With this comes the throwing into doubt of precepts and beliefs received from parents. Important aspects of the person's super–ego or conscience are thus thrown into disarray. A young person is likely, then, to feel depressed and agitated at the loss of this old inner stability and finds some solace in masturbation, self–love and withdrawal. This makes him acutely aware of himself, but his ability to think abstractly is also active so that he is prone to feel deeply about the world at large as well and his *effectiveness* in it. There is thus an acute sense of self interpenetrating with immediate awareness of the outside world. These are drawn together with wide, sweeping abstract ideas, some of which may or may not be infused with remorse and personal responsibility for parents as sufferers. Such perplexing experiences are likely to provoke a young person to search for, argue about and try to forge a consistent clarity of ideas of what he himself believes to be moral and just. It is most moving and heartening to hear a young person wrestling with issues like these. We are honoured listeners to the articulation and self–conscious clarification and hence strengthening of his own *personal conscience*. The most effective expression of this conscience working is perhaps in young people's political activity (Kewston, 1968).

It has already been stressed (particularly in Chapter 5, in the section on primitive guilt) that the roots of conscience probably lie in depressive feelings of concern about the childhood self's effect on others. If this sense of effect on others is depleted, then one dimension of *being realistic*, both morally and intellectually, is impaired. Only by facing experiences of guilt, without the depleting trickery of defences, can a person be in a position to *test out his effects on others*. Thus the capacity to be realistic, both socially and intellectually, seems to depend on the operation of conscience. Conscience is not an isolated moral issue; real working effectiveness depends upon its formation.

Summarizing: we can say that the mature young person is not so much under the sway of his imagined omnipotence nor of his fears of adult giants as is the child. Rather he experiences himself as having some effectiveness in a world of ordinary, interesting, good and bad, or 'whole' people. Likewise his internal voice of conscience is ordinary, measuring him as an ordinary person affecting and being affected by other ordinary, fallible human beings. Discriminations of the self, with their schizoid manifestations have worked with depressive concern to find a new sense of the self.

The ideals of conscience may be shared by others but need not be blindly received from them. Its final form might have a fundamental closeness to the beliefs of his parents, but if it has been hammered out by the young person himself, it becomes really his own and the source of his *integrity of character*. Kohlberg (1969) points out that individuals with such highly developed consciences are statistically in the minority. His is an American study but I suspect the results would be similar in other Western countries. In which case, it would highlight a gloomy fact that the number of people who have successfully worked out that fundamental issue of adolescence, finding their own conscience, are few. And even if one does when young, the anxieties of later years can easily muddle its clarity.

We have now passed through the acute internal questions of adolescence, the next chapter concerns itself with turning to the external world and testing the self in work contributions and sexual love.

FURTHER READING

1 Blos, P. (1979), *The Adolescent Passage* (New York, IUP). A technical psychoanlytic book, but Blos' work is standard reading for anyone specially concerned with this age.
2 Erikson, E. H. (1963), *Childhood and Society* (London: Triad/Paladin). A classic but particularly valuable about adolescence.
3 Fleming, C. M. C. (1967), *Adolescence: its Social Psychology*. A good general introductory text.
4 Laufer, M. C. (1974) *Adolescent Disturbance and Breakdown*. Brief and simple but not superficial.
5 Smith, A. (1970), *The Body* (Harmondsworth: Penguin). Particularly good about puberty.

CHAPTER 10

Work, Identity and Love

Testing the Self by Making Contributions to Others

Let us now turn to finding one's own *effectiveness* in the adult world by work and love. Naturally the sequence laid out in this book of adolescent internal crisis, followed by going out to work, and lastly falling in love, often does not take place in this neat order. All three can be concurrent and interweave. Even so, the order has some meaning, for the breakup of childhood patterns of obedience, the acute adolescent crisis, must precede the finding of a solid work identity if the latter is to be felt as freely chosen and deeply satisfying. Likewise we will argue that a person in our culture must have found an assurance about himself, to be alone and self–sufficient (through work particularly) if he is really to esteem himself as worthy of the lasting commitment of another's love and bringing up children.

The main stream of school, in Western culture as it stands at present, makes demands which have much in common for all pupils. But, when young people leave it, and if employment is available, each individual finds a place and kind of work which demands its own, specific form of skill and presents its own satisfactions and burdens. Young people find themselves in less–homogenous settings than school presented. I am afraid we cannot examine any of these specific work situations in detail here but there is ample literature on the subject (Carter, 1966; Terkel, 1985; White, 1975).

Testing ourselves in the adult world takes place largely through the *effective contributions* we make to others. We can do this in a variety of situations, in friendships, love, hobbies, sport, home, local activities and movements, and in *money earning work*.

We have arrived straight away into finance, hence inevitably to questions of economic policy. Whether we like it or not, we are in *political issues*. I, like most of you readers, live in a Western, more or less capitalist and market economy. Here, among other things, people are rewarded *differentially* by *money* for their contributions. These rewards can, of course, be grossly biased and hence hateful to many. In a market economy the financial rewards, for any given contribution, depend largely on the demand, or how much it is wanted by those who have got money to spend. The intrinsic difficulty, danger, pain, responsibility, or cleverness required to do a particular job are not directly rewarded, only the extent to which that service or product is wanted by those with wealth or influence.

How far this system, centrally based on the desire to maximalize financial profit, is the best one for people to live in is a very open question. In its pure form it is a greed system. However, it can give a freedom for individual initiative which is precious. But it often omits or devalues concern for others, relegating their care to a narrow, ineffective, private responsibility. Does not compassion warrant deep and wide social thought and political action at a community level? It must be of central public importance in close highly populated nations. Surely a culture valuing at least some socialist ideals can be richer and happier than one controlled by a greed morality. But what of its antithesis – has state socialism created happier cultures?

These are vastly important questions which cannot be investigated here even if I were competent to do so. We can only ask, and must go on asking these political questions whatever our work. But, even though this is not, of course, a political book, its subject still has a central function for political thought. The recipients of political systems are individual people. The final criterion of a system lies in the personal experiences, happy or otherwise, of individual conscious human beings. Political decisions can surely only validly be made with better knowledge of what brings individual happinesses and misery: and this book is, of course, an elementary essay in this quest.

As things stand at present *work* is often equated with *money earning activity*. Money earning is now almost universal in the world in both socialist and capitalist societies. It has the great advantage for every individual in that *self–earned money* gives *freedom to choose* one's satisfactions. However, money earning is a narrow definition of work. If instead we define it in terms of *intentionally making effective contributions* to others, then conscientious unpaid service to family, friends or the public would be felt as work. This accords well enough with general usage. What differentiates work from play is that one works for something more than immediate personal pleasure. Conscientious *discipline* is intrinsic to any work. It seems that work must be rewarded, not necessarily by money, but it is hard to conceive of a person going on disciplining himself for no recompense at all. Reward can often be from the knowledge of being useful, people's gratitude, pride of achievement, a little relief from fear of a tyrant, or even neurotic pleasure in suffering. But work without recompense is inconceivable.

When going out from school or college, a young person is faced with a variety of situations: friendships, hobbies, sport, home, loves and, with luck, paid work itself. He has to asses intuitively his contributions to and satisfactions from all these in shaping the style of his life and finding a satisfactory sense of himself. This is done only half–consciously, probably in a blundering fashion, and is naturally often seriously avoided. But making an assessment of our contributions is centrally important to growing up; it takes *courage*. If he fails, a person is left flitting in self–deception, with a very shaky sense of self–esteem and integrity. Furthermore, these contributions involve effects on others, so the scrutiny of *conscience* plays a central part in this process of finding self–satisfaction.

But work also entails loss of freedom, of time if nothing else; to gain some freedoms, with money for instance, *sacrifice* of others is inevitable. Making *compromises* is essential to working out any adult style of life. Here are a few examples of older people reflecting about this in conversation:

> I am a cowman of course, and it's good enough work. But politics is what stirs me. I wouldn't like it as a job, but it and the job go well enough together.

> In doing social work, I at last found I was happy. It was me. I couldn't imagine doing anything else. I relaxed, I suppose, in the evenings, but the middle of me was being a social worker.

> In those days football and motors were my life. I wanted to become a professional, but my dad said I must really help in the shop. Being in a village with no one else to do it I could see his

point, so I said, right I'll do it, but Saturdays must be mine for football, and it was agreed. Later, I ran a taxi from the shop and everyone hereabouts came to me if their cars wouldn't go.

As well as sacrificing time, any kind of work entails varying degrees of skill and differing forms of *discipline* to master anxiety and hence the use of defences. For instance, a dustman has to master his disgust, a spiderman vertigo, and a teacher has to master class discipline. A lorry driver must keep anxiety within bounds but maintain enough to be vigilant. And amongst many other controls, a nurse must restrain her fear of causing death while still having it available in her caring; she needs to be friendly but to check her sexiness; must be immaculately clean but not give way to disgust or nausea. You will by yourselves easily be able to specify what disciplines and defences are important in other sorts of job (Snelson & Erikson, 1980; Stevens, 1979; Terkel, 1985; White, 1975).

Repetition, Creativity and Responsibility

When a person settles into any job he will find varying degrees of *repetition*, together with opportunity for *creativity and responsibility*. Let us consider some of the personal strains and satisfactions involved in these. We are particularly thinking of paid work now.

Although automation is replacing much large scale repetitive work in conveyor belt assembly, all jobs entail much repetition, even the most creative. Even with robots and computers, in offices, factories and farms demand millions of man–hours of mechanical boredom allied to the minimum of creative interest.

The essence of repetitive work lies in speed of movements carried out time and again without variation. From the personal point of view, *choice of action* is excluded once the skill has been mastered. This seems to constitute the drudgery of repetitive work more than gross physical toil. Just as a child does, the adult mind seems to need exploration and variety for its satisfaction. When a task cannot be invested with imagination intolerable boredom is likely to set in.

The means by which those involved in repetitive work find

variety often have to be ingenious. Some people slightly alter their mode of carrying out a task from time to time. Music is another way. In listening, we can play in imagination while the body carries on with something else. Day–dreaming is a common outlet, so is talking, but this is only possible when the noise level is low. Using such means, many people find it easy to submit themselves to being mechanical for long periods. Perhaps the fact that anxiety is reduced to a minimum in repetition helps here. Incidently, Arendt (1969) has made a fundamental distinction between *labour*, which is essentially repetitive yet tied to the cycles of life, and *work* which is not. She suggests that when repetition is in the service of an essential cycle, as in agriculture, it is not necessarily intolerable but becomes so if alienated from such a cycle. Work, on the other hand, is enjoyable only with skill, choice and creativity.

Much work fortunately requires a great degree of choice. Thus a fitter has more choice as to how to go about a task than a conveyor–belt operator or copy typist. The greater the freedom of invention, the more *skilled* does the individual have to be. This entails self–disciplined sacrifice of idiosyncratic whims. For, like a mother with a newborn child, the worker has to submit himself to the limits of his tools and material. But, after perhaps years of self–discipline, he, his body and his materials have formed a sort of unity. Here we see a strange outcome of self discipline; this is a controlled loss of the boundary of the self; so that self, tools and materials become almost the same. A person's skill will then be a pleasure to perform and a beauty to watch, and his products of high quality. It is of interest that motor and spatial skill in the West tends to be placed on a lower plane than verbal intellect. In some parts of the East this has often not been so, the highest levels of enlightenment for instance have been regarded as only attainable through physical skill.

Art requires skill and more besides, it starts with a person becoming 'at one' with or the same as his object and materials. To be good he must lose himself in them. This recalls the loosening of the boundaries of the self referred to frequently in previous chapters. And then, as also mentioned, the creative artist tries to allow a personal abstract vision to arise which he reproduces. It is usually only after years of devoted practice, when he is master of himself and at one with his materials, that this is possible (Milner, 1957).

167

Skill and art can, of course, be present in anyone's everyday work paid or not. It is often tainted, however, because most of us cannot be bothered, or have not the time to develop mastery, so we have to make do with unsatisfactory, botched–up jobs. It is of interest to note that skill and truthfulness may have much in common. A skilled worker knows truthfully about his materials and their functions. A botcher makes much *pretence* of knowledge, and deceit can then, of course, be towards himself as much as to any customer.

A good artist seems to have a particular form of deep conscience which works in a way that is baffling to many people. He feels a fundamental responsibility to his art, to purify his abstract vision and its communication to others. He is often not so deeply concerned about the everyday effect on other people and can even be indolent and thoughtless about common sociability.

For most people, however, responsibility means concern for others, directly or indirectly. Let us examine this, for within responsibility lies, I think, a keynote of *adult maturity*.

Both repetition and creativity were exercized in school, so they will not be new to the working adult. But full *responsibility* for a task upon which other people depend will be relatively new.

Responsibility for a task means being entrusted with its successful completion. You will remember how the infant first needed to trust his environment and then began to trust himself. With this he began to be responsible for his own functioning. By adulthood, however, other people will expect to put their trust in him. They will become like infants, as it were, in his hands during the period of the task he has to perform for them. Thus passengers need to trust their driver, housewives their plumber, or patients their nurse.

The individual's sense of responsibility involves knowledge of the task to be carried out. It also requires other functions. His *conscience* must be operating so that he recognizes the existence of others and feels *concern* about his effect on their well–being. This operates to inhibit impulsiveness or withdrawal from the reality of the task. It requires *continuity*, sticking at the task no matter what personal impulses arise. Lastly, it involves *anticipation* of possible trouble and disasters. The responsible person must be able to tolerate anxiety about possible *faults*, both in himself and in the situation in hand. Only then can he realistically watch over it. The

168

exercise of responsibility has great satisfactions. The individual is master of the situation and is also being *parental*. This involves burdens of personal anxiety and guilt. To be responsible, a person must use his feelings without being overwhelmed by them or being carried away into egocentric fantasy.

Each task provokes its own particular anxieties, that some people can bear while others cannot. For instance, an individual may find that anxiety about heights is too much to contain, so he would not make a good airline pilot. On the other hand, his anxiety about vomiting, death and damaging others may easily be under control, so that being a doctor would be quite tolerable.

Many tasks involve no direct responsibility for other people's well–being. But all jobs involve it indirectly. Thus a man making spare parts for washing–machines is not directly responsible for other people's lives. But if he fails in his job, a housewife may be affected months later and miles away.

This description of responsibility will probably make it sound pedantic and dull. Its essence does carry a necessary solemnity and sameness which is dull. It entails conserving and is thus, as it were, conservative. And in many ways it is antithetical to romantic exploration and personal creativity.

It is perhaps the loss of romantic freedom which makes responsibility loathsome to many people, young and old. They seem to associate it with dullness and hence deadness. Perhaps they are frightened too, not only of depression, but also of possible guilt. Being responsible is also burdensome because one often has to direct other people who are likely to be cussed and revolt against orders. It is perhaps these factors which make heavily responsible people feel that they deserve high recompense both in large salaries and luxurious perks.

In addition, a person apparently weighed by responsibility can also, often quite unconsciously, gratify his yearning for omni-potence by playing covert ruthless games. For responsibility does carry with it power over other people's lives. Politicians particularly, are notoriously prone to indulge in ruthless power games even though they frown with the burden of their immense responsibilities.

But some people can carry responsibility with a quiet natural-ness. It need not be flashy but can be deeply enjoyable. Its steady working, with self–scrutiny to watch out for insidious ruthless games, is essential to the good working of our, or any, society. It is a 'must' in parenthood unless we are unconcerned about shattering the growth of those we have brought into the world.

Self–scrutinizing responsibility is the keystone of helping professions: here it needs cherishing because it can so easily be swamped by histrionic or paranoid games play. Perhaps these professions could more forthrightly lead in responsibly cultivating a climate of understanding of individual predicaments in life.

However, if 'adolescing', with its regression, romance and madness, is not kept alive throughout our lives then we neither have the means to change ourselves, nor can we provide the environment for others to change. We become either very dull, shrunken people unable to face crisis, or arthritic martinets, self–satisfied with power and unable to be either humble or flexible.

Identity

We have just been considering the young person, if he is lucky, facing the situations of: paid work, friends, family and leisure. Here each specific activity can be considered as a *role*. At any one time a person has a multiplicity of these. When, and if, he has found a *personal patterning of these roles*, and where the emotionally charged contradictions between them have been more or less disentangled, and which then *together give him inner satisfactions, he then usually experiences a sense of wholeness and well–being*. This integration is often referred to as *adult identity*.

This identity, though delimited by forces of one's culture, as well as inheritance, is essentially idiosyncratic and private to each individual (Erikson, 1963, 1968; Lichtenstein, 1977; Lynd, 1958; Smelser & Erikson, 1980). We have been slowly leading up to this concept of identity particularly when we have been stressing that the honest scrutiny of our conscience is essential in finding *integrity*. For it is by self–scrutiny that we assess and balance the enjoyments of our own life with those of others. We have stressed many times that this conscience seems to be born in the experiences of infancy and is readily shattered, distorted and impoverished. It is for this reason, perhaps beyond all others, that psychoanalytic therapists are often so concerned about the early years of life.

170

In previous chapters we have also been stressing the importance of intellect, particularly its logical activity, in emotional development. We have for instance suggested that *emotions may be viewed as thoughts that are not yet fully formed.*

We can now see that the desire to be logical, with stress on non–contradiction and hence consistency, plays its part in the integration of the self's identity. We may always have to negotiate conflicts and contradictions in our life roles but, without a truthful awareness of them, we cannot realistically feel we have integrity.

Erikson's idea of identity, though of quite recent origin in systematic thought, has now become part of our everyday language. We thus hear of work identity, sexual identity, class identity, racial identity and national identity. Because these concepts have entered into everyday thought so quickly and powerfully they must reach deeply into us and deserve most serious consideration, argument and books of their own. We can only draw attention to them here and point out that they are slightly different from the overall individual self–identity we have just introduced. Sexual, racial, national, class and work identities refer to the individual's feeling of belonging to certain groups as distinct from others. A person must experience a multiplicity of such identities for he belongs to several of these groups at once. They are as it were, sub–identities of a person's overall feeling of being himself, his self–identity.

Residues of Childhood

Even though a person assumes the responsibilities of being an adult, and his childhood–dependent attitudes slip into the background, this does not mean that they have died within him. The childish parts of himself must remain active, often unconsciously. When unassimilated into one's present life their insistent repetition may make one prone to *neurotic symptoms* of one kind or another. We cannot discuss this fully here but a little introspection will probably make it plain to the reader that there are urges in himself which continue more or less unchanged from his own childhood. Here are a couple more illustrations.

A young boy was brought up in the country. His father ran a

progressive boarding school which was regarded with some suspicion by the local people. The boy's mother cared little about this, and ran her family with a back–to–nature philosophy. Her pigs, goats, hand–loom weaving and home–made pots were unusual and were laughed about in the neighbourhood. The young boy grew up isolated from the local children, though he went to school with them. He was baffled by them and envied their conventional toys and games, so that he rarely felt at ease to join them. When he grew up he became a schoolmaster like his parents. His resemblance to them seemed to end there, for he went to work in a city and was more conventional in his ways. He was respected and very well liked, so that he quickly became head of a department. However, as soon as he cast off his professional mantle he was crippled with shyness. He could not bring himself to be at all intimate with others in the neighbourhood. When not at work, people outside his family seemed like a race apart, who would look askance at him just as they had done years before.

A boy, many years ago, was one of the elder of many children. He was very fond of his mother, but each year brought another birth, so that he had but little of her company. He took this stoically, and in fact became particularly admired by his family for his bravery both at home and in childhood exploits. His mother died when he was ten and his father struggled on alone to bring up the family. Two years later the young boy said he wanted to relieve his father of one mouth to feed and was determined to join the navy. His father gratefully accepted this further demonstration of bravery.

When he was eighteen, after six years at sea he fell ill and was invalided home from the East. In the ship on the way back he seems to have gone through an acute adolescent crisis. He became aware that he had no future in the navy, and could see no way of life that would fulfil him. He wondered whether to throw himself overboard and finish it all. Then a 'radiant orb was suspended in his mind's eye'. He exclaimed, 'I will be a hero and, confiding in providence, will brave every danger'.

Later in adult lie this is just what he did. As Lord Nelson he became the brave hero not only to his family, but to England at large and to succeeding generations of schoolboys.

In both these examples we can see elements in adult identity which are from the past. These are both creative and neurotic at the same time. For instance Nelson's heroism was close to suicide but it was inspiring heroism nonetheless. We can wonder what makes one overbalance the other.

Compromise and Adult Identity

Throughout adult life, demands to *conform* to other people's requirements are continually impressing themselves upon the individual. At the same time, he has his own inner wishes, so that conflict becomes inevitable. If a person is to maintain an integrity of his personality, he has to make *compromises* between his own wishes, his conscience, and demands of the outside world. It seems that integrity is lost when chronic *splitting* with *denial* of emotions occurs, so that one set of interests obliterate others. Fanaticism is one instance of this, and a person preoccupied only with himself is another. So, too, would be someone who surrenders himself totally to others' demands. For example, compare these two men, both bank managers.

Mr A was highly esteemed by his superiors and customers. However, his whole life was consumed by his bank. He had no hobbies. People were invited to supper on the basis of their business connections. His family life was invaded by ideas and strictures about the bank. He would say to his wife, 'You mustn't wear trousers in the front garden because customers pass'. Or, 'You should order your groceries from Jones's because they bank with us', and so on. In submitting to the bank he was unable to enjoy either his own bodily functioning or the inner lives of his wife and family. He repeatedly fell sick and his wife, failing to make an impression on him by ordinary means, soon found herself turning into a chronically hysterical individual. Although esteemed by many, he had failed to go through the agony of making an integrating compromise between the various demands of his situation.

Mr B was actively interested in athletics, music and literature when young, but went into the bank to earn a living. When quite an old man he said, 'I often thought I would leave the

173

bank and do something that directly interested me more, but I fell in love and decided that getting married was the more worth while. And to do this I had to stay in the bank.' However, he did not forget the expression of his inner life. He continued to be a reader, talker and musician in his leisure time. 'The bank was satisfactory in a way. It was secure, and looking after people's affairs was interesting. But everyone knew my heart was in other places as well, and because I wasn't married to the bank I didn't get far up the ladder.' Here we see evidence of making compromises, but not of losing self–integrity.

We see in Mr A a common form of fanaticism though it is not usually graced by this term. He was obsessed by ambition and comformity to banking. Mr B on the other hand, tried to evaluate the various aspects of his life and sought to compromise between them. Because of this he may have suffered the fate of being thought ordinary, but no one would say he was a nonentity, and many were glad to have known him. Mr A *compromised himself*, he lost his integrity by denying so many parts of himself. Mr B on the other hand made *compromises*. The course through adult life seems to be a continuous compromise between conflicting issues. As new situations arise, so the individual is called upon to develop new patterns of thought, to find new compromises which will maintain his own and others' integrity. It seems that many people muddle these two meanings of compromise. They think a person who makes compromises is a weak or flawed character. This is plain foolishness, of course, and tacitly gives assent to fanaticism.

Continuity of Adult Identity

Some writers seem to think that once an adult identity has been found, in the mid–twenties say, this should now normally remain set through life. Even Erikson, to whom we are indebted for introducing the concept, often seems to imply it. It may have been true when life was short and societies relatively static. But even a cursory glance at lives today points to the many shifts or roles, often of a fundamental kind, that individuals have to develop through life. A woman, for instance may leave school, genuinely find her independence through being a secretary, then fall in love, marry and forget all about her previous work in engrossing herself in mothering. When the children go to school and grow away from her, she may find herself dissatisfied with married dependency and seek a new, paying career which offers more personal creative satisfactions than a secretarial office. So she becomes a student again, trains and gives herself a pattern of responsibilities she has not known before.

All these require change and development of adult identity, and each change is likely to be a life crisis of greater, or less proportions. This being so, the individual must expect to proceed through a break-up of old patterns, loosening into regression and testing new modes of competence of a similar form to that of 'adolescing'. These later crises can often be even more searching than the teenage shifts. It is for this reason that so much time was spent on adolescence, apart from the central core of puberty, its form can be repeated, with modifications, at any time of life.

Now let us turn to the other fundamental of young adulthood, in our culture in the late twentieth century at least; falling in love.

Falling in Love

In our society, when the subject of sexual love arises the question of marriage is not far away. But for the time being we shall ignore marriage.

Turning our minds back to previous chapters, the whole sequence of the history of erotic development through life will be recalled. First came the baby's erotic love with his mother and the necessity for harmony and synchrony between them for healthy physical and mental growth. Then came frustrations of this love, easily evoking experiences of extreme distress and hence mad rage. At this time, also, intellectual events were readily accompanied by gross physical reactions. Thus the whole complex web of meaningful enjoyment, puzzlements, frustrations, rage and despair that a child felt, with his mother in particular, was interwoven with his eroticism. In other words, meaningful love and eroticism have roots in the mother–child relationship and their bodies together. This was continued in conflicts over social and bodily control in the second year of life. And then, though less obviously, the joys and conflicts of the child about his parents seemed to be sensitively focused upon genital urges and fantasy. This was epitomized by the Oedipus complex. With growth into relative independence and going out into the company of other children, the child seemed to be less erotically attached to his parents and assumed a certain detached dignity; latency had set in. Though he was easily arousable and sexually curious, there were probably few major changes of sexual feeling until puberty. At this point the arousal of the genitals and body growth insist that a child is a sexual being like his parents. This acted as the trigger to the adolescent sequence which, if successful, culminated in the ending of dependent expectations about parents.

We have noted that the first directly sexual feature of an adolescent's turning away from parents is crushes on other people. These tend to have a disconnected and primitive all–or–nothing quality about them. They are passions for *parts of people* or *abstracted* physical characteristics: for instance, firm mouth, mellow lips, copious breasts, or the way a head is held. These are physical attributes, but they usually also symbolize *personal characteristics*. For instance, a firm mouth reminds the adoring one of 'determination', mellow lips suggest 'sensuousness', copious breasts evoke ideas of 'generosity', and so on. They are *physical metaphors*, rather like *dream symbols* in which physical attributes also stand for more psychological functions. This contention is supported by the fact that physical attributes are not the only things adolescents fall in love with. Psychological characteristics of a person are often quite explicitly felt as objects of passion. Here are one or two examples of young people's reflexions about their loves.

I am in love with Jacques more than anyone, it is I think because he, like I, is half in love with death. (fifteen–year–old)

I love Dean, he is nothing but pure kindness. (fourteen–year–old)

I only fall for hard men. I think it is because they are the only ones who could stand my violent temper. (seventeen–year–old)

Why are these abstracted parts of a person's character fallen for with such deep personal meaningfulness? The girl who gave the last example gives a clue. She sensed she needed someone to control the rages she could not manage herself. She *fell in love*

176

with a characteristic she felt *deficient in herself*. This tallies well with the age–old saying of, 'Falling in love with your other half'. This stems from Greek mythology, which held that man had originally been happily whole, but later one–half of the soul had become separated from the other. Thus mortal men were always incomplete, searching for their other half.

If at least one aspect of falling in love is towards a characteristic we feel deficient in, then we would expect that a person would feel extremely humble compared to his adored one. Idealization would be a facet of love. This is certainly the case. You will remember how love and idealization loosens one's conscious experience of emotions so that awareness of limits or boundaries between things become emotionally blurred. One feels towards infinity. The loved one is all goodness and beauty and the world itself is infused with beauty too. So the loved one is beauty, which at the same time is *in* all the world. It is, of course 'quite mad', this 'infinitizing' of things. But if one has, in some sense, experienced it, if only once, life seems to have been worth living.

However this sort of level of loving is experienced in an all–or–nothing way; it is primitive and readily ruled by splitting. This being so, we would expect even slight misunderstandings to lead a humble lover to switch to violent hatred at being betrayed. This is certainly likely to happen with immature crushes. They are *unstable*, what is more, since idealization of the loved one tends to *deplete* self–esteem; a lover readily becomes explosively envious. Only when the *idealization is reciprocated* does this self–depletion receive a counterbalance of idealization of the self.

Obviously this is not a complete unravelling or understanding of the crush. The girl quoted above who was in love with Jacques, for instance, said she loved him because he was *like herself* 'half in love with death'. Here she had found a comrade in her suffering, the two, alienated from the rest of humanity as they saw it, could understand and comfort each other like babes in the wood.

These are just two positive lines of inquiry into the riddle of why one person falls in love with another. It is certainly not exhaustive but is, I hope, at least a start.

At this stage we have simply discriminated some types of characteristics that may be fallen in love with, and of course they could belong to a person of either sex. In 'adolescing', apart from social pressures, movement of love for a person of one sex to love

177

for the other is usually fluid, at least at the level of fantasy. But we have so far described only the crush, the first stage of being in love. Let us proceed to the second where the fantasy crush generates a real two–person relationship. We shall consider homosexual choice first, and then heterosexuality (Geer, 1984; Masters & Johnson, 1966; Schofield, 1973; Shor and Sanville, 1978; Tiefer, 1983).

Male Homosexual Love
(Riutenbeck, 1973; Rosen, 1979; West, 1960)

You will remember that according to our definition in the last chapter homosexuality falls into the category of a perversion. That is because the man's genital does not find orgasm within the genital of a female partner; a part of another man's body is used instead.

Arguing on from this point; we may assume that humans have biological urges to reproduce like all other animals. Thus we have biologically determined urges to have heterosexual intercourse. This is *avoided* in homosexuality and, as the redirection is against the strength of a biological drive, there must be a *revulsion* against the heterosexual act; it must also be profound. What is more; the revolt, either in terror or disgust, must be primarily against the idea of the penis in relation to a woman's genitals. Usually almost any physical contact with a woman is avoided in homosexuality, but the main dislike is the idea of the penis within a vagina.

Women themselves *as people* are not necessarily disliked by homosexual men; it is the *sexual arousal* that is. Here, however, there are usually fears of the most fantastic kind. We cannot go into this in detail but it is worth noting that *paranoid anxieties* are usually not far away. Let us restress that the biological reproductive urge, which is very powerful indeed in all mammals, has been diverted. Something psychologically very intense must be happening.

Close to the revulsion against 'penis in vagina' very often goes an *idealization of the penis*. This may involve a desire for a man to admire his own penis or it may be to adore another's. It is not universal but penis adoration is an important aspect of many 'gay' cultures.

178

Close to this is often a profound enjoyment of physical make–belief. This involves the playing of 'male' and 'female' roles in sexual partnerships with the anus standing in for the vagina as often as not. This again is not universal in homosexuality but it does point the way to the importance of *anal excitement and fantasy* in much male homosexuality. If we remember how discipline and control are intimately related to primitive anal fantasies, then we can recognize that homosexuality also seems to involve a revolt against certain levels of discipline. This is perhaps confirmed by the observation that homosexuality itself is a *revolt* against 'ordinary' biological ways.

None of this will be fully convincing to the doubter but I hope it is enough to stimulate thought. Remember too that heterosexual loves themelves can also be full of perversities and paranoias. We are not here trying to extoll the healthy virtues of heterosexuality at the expense of homosexuality. But we should not specially plead for homosexuality either.

To get the flavour of a homosexual love here is a brief description of such a choice. It is anecdotal, but may suggest wider lines of inquiry.

A young man had, since earliest childhood, been preoccupied with the necessity of pleasing people. His father said he seemed very happy and very clever when a little boy. If there was anything wrong it was that he was too good! He got on well with his father who was sensible and quiet. He doted on his mother who histrionically lavished adoration on him. This was matched by the company of many old ladies whom his mother also looked after. He was precocious intellectually and hence quickly became something of a teacher's pet at school. Perhaps because of this he tended to be very frightened of the jealousy of the other children and, being small, had no means to assert himself physically. He yearned to please them. Going into his teens he became readily acceptable as a clever boy, but was still obsessed with the need to please. At this time he began to consciously dislike his mother, feeling that she overwhelmed or ate him up. He also began to be aware of how ruthless he could be and noticed that he often dreamt of being a despot, like Caligula for instance. But he was in fact effectively unassertive, and awareness of his ruthlessness doubled his wish to be nice

179

and liked. In his late teens he tried a lightning affair with a girl but was thrown into a chaos by her flighty histrionics and discreetly retired from the encounter. Therefore he went on liking women, and had many as friends, but sexually he kept his distance.

Then he fell passionately in love with a brash, ruthless young bisexual man who was apparently socially more outstanding and successful than he was. The Caligula of his dreams had been found, in actuality. The young man was enthralled. The ruthless man seemed to find kindly innocence appealing, so they fell into a passionate affair. The ruthless man, however, soon tired of this and found another *amour*, leaving the young man hurt and bereft, determined to give himself to gentler lovers next time.

In this we see many elements. Not only is there a *revolt* against 'good' practices, together with a dread of being *overwhelmed* by a woman's excitability, but also a *hunger for the character of the loved one*. In overt homosexuality this hunger for another's character often seems to take a physical form. It is wanted physically, even to incorporating the loved one's body, and penis in particular. This is the *idealization of the penis* which we have just mentioned. It may be wanted inside and this can be achieved in oral or anal intercourse. Or it may be wanted as something to fondle and worship in mutual masturbation. All of these are, of course, common in homosexual intercourse.

Our single example cannot validly be extended to general conclusions about homosexuality. But I think you will find the features mentioned here are quite common.

Lastly, remember that not all love between men is perversely homosexual. Men can love each other as people very passionately, there may often be sexual feelings involved, without there being any exclusion of each other's heterosexual loves. Thus two happily married men can be deeply fond of each other without any loss of love for their wives. Though every one, no doubt, has some homosexual urges, the noun homosexuality is perhaps best confined to refer to a man's sexual desire for another man to the exclusion of women.

We have not yet mentioned *bisexuality* except of course to suggest that everyone probably has sexual feelings about both sexes. Bisexuality proper refers to those who enjoy actual sexual intercourse with members of both sexes. We will not go into this in detail here. We can point out however that bisexuality does seem to instance how some people can *split* their lives into mutually exclusive, contradictory even, roles. They can at one time be passionately homosexual and put this aside quite quickly to turn to a woman. Whether this entails falsity must remain an open question here.

Female Homosexuality or Lesbian Love
(Chassequet Smirgel, 1980; Tieffer, 1983)

Lesbian loves usually have, it appears, rather different psycho-dynamics. There must be a similar revulsion against genital arousal. This time of course it is against the penis relating to a woman's vagina. Some lesbian women get on well with men, some enjoy 'being a man' in male company. Their love of women seems often to be with themselves in the role of a man. The function of physical make belief is clear enough here.

Many other women profoundly loathe not only penises but other aspects of maleness as well. Here the penis seems to be a physical symbol of all 'macho' qualities against which women must cling together in mutual love for protection. Here the paranoid elements are applying not only to genital arousal but to masculinity as an abstract idea. However, as we have noted, there are many women who can like men as people and whose loathing of maleness is apparently confined to the genitals and physical contact. Another common feature is the positive desire to be mothered sexually. Men as such are not so much loathed but objects of active indifference.

When a woman loves another sexually she often also hates her. This hatred can be denied and projected violently onto the idea of men so that the woman can say to another as it were: 'we don't hate each other, it is men we loathe.' We are then close to the anti–macho relationship mentioned in the first paragraph.

Here is one brief description of some aspects of a lesbian woman's life.

A young woman had been brushed off, when a little girl, by her mother who inordinately doted on her older brother. He was a rather soppy cry baby, bitterly envied and yet held in contempt by his sister. Her father was rather distant but loved by the little girl, however he was careless about the home and was openly denigrated by her mother.

When grown up, she liked and was liked by both men and women. But when close to men she found herself overwhelmed by hate and contempt for them. Being unsure of herself in intimate relations she began to feel very lonely and yearned to be fussed by a woman, as her mother positively had not. She met a colleague who was openly contemptuous of men but warm and affectionate to women and also wanted to be soothed by female affection. They naturally fell for each other and lived together for a long time. However it was not entirely satisfactory. One young woman loved children and wanted a baby, the other was hungry for the company of women with strength, definiteness and power so that neither was very fulfilled.

Here we have a pattern that is different from the male homosexual; there is *envy and contempt for sexual men* and yearning to be *fussed over in a motherly way* by women. This again is only an anecdote, but it shows themes which can be quite common in lesbianism (Cameron, 1963; Rosen, 1964; West, 1960).

Remember also that as with men not all women's love for each other is lesbian. Passionate affection can continue between them without the exclusion of heterosexual loves.

Values and Difficulties in Lesbian and Homosexual Relationships

At this point it must be emphasized that we have described only a few of the vast multiplicity of feelings and fantasy that are set loose in sexual arousal of any sort. Many involve projections of the self into another, or fusions of another into the self. Others involve, as we have noted, fantasies of being of the opposite sex to what one physically is. These fantasies are thought of as repugnant by many people and hence to be condemned. Such condemnation, often of a vehement kind, seems to be driven by the protester's own hidden fears of perversity and usually has little real substance.

However, I think this should not blind us to the possible intrinsic *instability* that comes with obliterating one's heterosexuality. This instability, at least when held in fantasy and not acted upon, has great value; the shifting of identification of self from one sex to another in imagination, for instance, is essential for sympathy with the other sex and for creative breadth of vision. It is also intrinsic to certain areas of learning and work. For instance, if a man is to learn how to nurse babies, or if a woman is to model herself on male teachers to develop competence in her work, then both need to enjoy identifications with the opposite sex. It is, I think, one of the major social achievements of recent years that prejudice against this cross–sex identification has begun to relax. Among other things cross–sex identification enriches heterosexuality. A woman for instance can enjoy identifying with a man when she has a real man's penis inside her and vice versa.

However when actual homosexual or lesbian relationships occur, certain *impoverishments* do seem to be inevitable. This because the biological heterosexual relationship is obliterated, denied or otherwise revolted against. The young man we quoted, for instance, did not find a relaxed use of those feelings which made up his ruthlessness when he fell for a ruthless young man, he simply got hurt. He was excluded from using his ruthlessness *mixed* with warmer feelings to give him firm thrustfulness, either in a sexual relationship or in other affairs. A large part of him had to play the innocent. And the young woman could not satisfy her yearning to be engrossed with a baby by lesbian intercourse. This raises the question as to how heterosexuality compares, so let us now turn to this to continue the argument.

Heterosexual Love
(Geer, 1984; Masters & Johnson, 1966; Scharff, 1982; Schofield, 1976; Tiefer, 1983; Tunnadine, 1983)

As in homosexuality, perverse fantasy is usually embedded in heterosexual love, but there is something else as well. We have just noted that in both male and female homosexuality there seem to be severe persecutory and paranoid feelings about arousal with the *opposite* sex. These persecutory feelings of fear and hate are strong enough to swamp and then freeze heterosexual urges. A person who can turn to heterosexuality is not ruled in this way by paranoid fears about sexual arousal. The fears must be either milder or strongly defended against. Remember heterosexual performance is no guarantee of health at all.

Whether it be a homosexual or heterosexual love, a young person, taking the momentous step from pubertal crush to an actual approach to a boy- or girl-friend, opens up the possibility of sexual reciprocity. We referred to reciprocity in the early chapters when describing those mutual contributions to each other's self–esteem which can occur between parent and child. In these younger days such reciprocity was in a one–sided context, the parent really held power, now it can be more equal. As we noted earlier, the idealizing lover can be loved back so that he not only feels humble but, by being idealized in return, he is made to feel wonderful as well. He probably comes to feel, and be, more vital and colourful than he was before.

Where this reciprocity takes place with someone of the opposite sex the two, obviously enough, physically fit like jigsaw–puzzle pieces. The man's physical ecstasy, centred on his penis with its particular form, fits with and affirms the woman's ecstasy, centred on her vagina with its different shape. As their body shapes complement each other, so too does much of their fantasy which is closely tied to body shape and impulse. And, since thought patterns are fed by fantasy, so there is the likelihood that more intellectual functions can work in harmony too.

It is this mutual body acceptance that can give a blessed inner stability and peaceful harmony to young lovers. At best, unlike homosexuality, there are no loose ends. There are, as it were, no penises with nowhere to go, no unused vaginas or breasts that will never give milk. This is important, if only because it is neglect of these parts of the body that can give rise to violent, painful, even uncontrollable paranoia and destructiveness. These neglected parts are somewhere felt as belonging to the self, even if unconsciously, and when a self feels neglected it is hurt and easily becomes destructive. It is the harmony or synchrony between two people of the opposite sex, rooted in physical intimacy without loose ends, that homosexuality and lesbianism seem to miss.

184

Having said this, it is as well to recognize that heterosexuality can itself become a chronically defensive trick. For instance, it may be used as a flight from depression while at the same time being swanked about as the 'height of normality' to less fortunate creatures. Naturally this is a subtle but unpleasant form of cruelty. It can be a make–believe against fears of homosexuality, passivity, impotence or frigidity, and all sorts of other anxieties.

Remember too that many homosexual or lesbian partners are much happier, fruitful and creative with their partners than with no–one at all or with an incompatible member of the opposite sex.

Heterosexual lovers are usually not satisfied to rest with physical harmony which, even though it may give rise to intellectual enjoyment as well, is often transient unless there is a deep, reciprocal appreciation of *character* as well. When this appreciation occurs, for instance, a person may have some ability which he believes in but is unsure of, and deeply wishes for it to be affirmed by a lover. If it is not noticed a person can feel just as hurt as if his whole soul and body has been spurned. The 'chemistry' of *whole characters in interaction* is as important in love as that of bodies. This interaction starts with idealizing crushes which, as mentioned, seem to involve a falling in love with aspects of character. This idealization by itself can be full of contradictions and very instable. But for lasting enjoyment something of the following process must also, perhaps, take place in reciprocity.

A lover may idealize and want some characteristics for himself but then sense that if he had them they would be antipathetic to him as a whole person. When this occurs, he may be glad of a partner with these ideal attributes, so that he can possess them at one remove.

Here is a simple example. The characteristic used is a part of the body, but it could apply to any characteristic of bodily or mental functioning. A man may particularly idealize and dote upon his girl–friend's delicate, expressive hands. It might be that he simply wants expressive hands like hers for himself. He may yearn to be a beautiful musician, but feels he could never be one because of lack of dexterity. In such a case he might simply feel his own lack, and his loving idealization may turn to corrosive envy. In another case the young man might equally like to have such delicate hands. But, at the same time, he is proud of his own heavy, muscular hands, and delicate and heavy hands just cannot go together in the same person. So as a *whole person* he does not want delicate hands. But he is very glad that he can 'possess' them at one removed in his loved one.

Likewise his girl might want, in part of herself, to have strong hands, and yet be proud of her own delicate ones. She reciprocally will thus dote upon his strong hands, while being pleased by her lover's appreciation of hers. Both partners are happy. This is a trivial example but the notion can be applied to many more complex characteristics.

We have argued here that, because of complementary sexual body shapes, heterosexuality is *potentially* the most satisfying stable and harmonious form of love.

However, I must add that heterosexual falling in love may be necessary in harmonious love, but, it is not sufficient. Simple *liking* and *respect* for each other seems to be just as, even more important in the long term than the more romantic aspects of loving. But let us continue with this in the next chapter.

Lastly, something of little recognized importance. Though kindly heterosexuality is of inestimable value in a couple's intimate life, it is vitally necessary in a much wider way as well. *Inhibited genital sexual feelings* underlie the ordinary conversations of men and women everywhere, in work as well as leisure. I am not referring to trivial flirtatiousness which can be full of innuendo, but to dignified, respectful *charm*. With this, in a quiet way, men and women, married or single, can enjoy each other fully as human beings including their sex. There need be no suggestion of physical lust or indecency. It is sanity at its simplest and best. When recognized, it is perhaps the surest of guards against madness of any kind.

FURTHER READING

1 Carter, M. (1966), *Into Work* (Harmondsworth: Penguin). A general text on this subject.
2 Davies, D. R. and Shackleton, V. J. (1975), *Psychology and Work* (London: Methuen). One of the 'Essential Psychology' series giving a straightforward introduction to this subject from the academic point of view.
3 Erikson, E. H. (1968), *Identity, Youth and Crisis* (London: Faber). A

series of papers, particularly about youth and the concept of identity.

4 Fisher, S. (1973), *Understanding the Female Orgasm* (Harmonds-worth: Penguin). Arising out of the work of Masters and Johnson but with considerable discussion of the interpersonal issues involved. I wish there was a companion to this about male sexuality.

5 Fromm, E. (1957), *The Art of Loving* (London: Allen & Unwin). A very popular book. Some readers will find it rather solemn and old–fashioned, but this is perhaps its great virtue.

6 Scharff, D. E. (1982), *The Sexual Relationship* (London: Routledge & Kegan Paul).

7 Shor, J. and Sanville, J. (1978), *Illusions in Loving* (Harmondsworth: Penguin).

8 Storr, A. (1960), *The Integrity of the Personality* (Harmondsworth: Penguin). Although some years old, this is still one of the best introductions to psychodynamics generally and to identity and integrity in particular.

CHAPTER 11

Partnership and Marriage

In this chapter we will discuss any kind of *living–together* sexual partnership where both people have agreed upon a *continuity* of relationship, whether explicitly unto death or not. We will be concerned primarily with heterosexual partners, but a reader can easily extrapolate those aspects which could equally apply to homosexual partners living together.

As the state of legal marriage is only one sub–category of partnership, we shall be asking whether it has in itself any value or whether it is an outmoded relic.

There is no doubt that in very recent years, in Britain at least, it has become perfectly respectable among many people, for a couple to live sexually together without being married. A few years ago this was outrageous to all but a few 'bohemians'. Now it often seems even more moral to live together unmarried than to rush to a wedding in an impulsive romantic whirl.

Whichever way they do it, when two people commit themselves to live together, for more than a 'holiday', for an unspecified time (which can easily become emotionally infinite), it is a very real personal psychological event.

The last chapter spelt out what is obvious: that when two people sexually fall in love, whether they go to bed together or not, there is the opportunity for expansion to their lives with a harmonious stability that has hardly been known before.

When reciprocal appreciation is found much more than sexual feelings are released. Because the reciprocity is sexual it is close to dreaming and is thus rich in fantasy, primitive, deep and very personal; but intimate friendship can be found as well. The intellect also seems to become more vital; our chapters on early infancy suggested that intellect is born of personal body fantasy; this seems to repeat itself when adult mutual love is found. And, with steady appreciation, the couple's shame at exposure of personal absurdities seems often to melt away. The comfort of

188

companionship has been found, defensiveness with its egocentric arrogance seems to matter no more, at least when the two are alone together. With humble selflessness in the ascendant, one's conscience often glows with a love for all creation. How long such wonders last will, of course, be an open question.

Because of the mental closeness of lovers, communication comes with quick intuition. The two can find themselves growing into *being a team*, moving efficiently together as one unit to solve problems.

It is a wonderful, stabilizing magic, but also explosive. Lovers live, whether they think so or not, in a social matrix. Their relationships with others are affected. Work, other friendships and family ties may become dimmed. What is important to one lover may be of no interest to the other. Habits of one may jar on the other. Sooner or later some sacrifice of personal idiosyncracy, which has been freely enjoyed in solitude, must be made in a partnership. But this is a partnership where primitive feeling and fantasy, stemming from those of a little child, have been loosened. This primitiveness entails splits into all–or–nothing feeling, things can be loved with consuming adoration or violently hated. Add to this the loosening of *self–object discrimination* so painfully developed through earlier life, and which may arouse anxieties of a psychotic intensity, we can see that the outcome can be very explosive indeed. Lovers' quarrels are usually comic to an outsider, but to the participants, who are defending their own precious selves against threatened invasion, it is not funny at all. Only later, if they can look at themselves with detachment, will they be able to see the humour of it. I would suggest that the lovers' quarrel encapsulates every facet of an individual's *early problem of social control*. Thus every form of *personal habit*, from meticulousness and untidiness to more general cares comes under attack, sometimes with lavatorial obscenity, but more often with simple disgust and contempt. On the positive side, the healing nature of laughter after a quarrel shows how precious is the ability to detach from the strident defence of self and see oneself as others might.

This *seeing of the self from another's point of view* seems crucial in any partnership, most of all between lovers. This moving to view things from *another position* has been a function of intellectual maturity as we have often stressed in past pages.

Being able to move from one point of view to another without losing the recognition of other views is vital for interpersonal maturity too. Is not this ability to move positions without contradicting or betraying previous ones the prime constituent of a person who can be trusted? A person who is trusted is also liked and respected. It is the mark of a person of integrity. This must be essential for people living together.

Courtship and Living Together

Traditional courtship must be seen as developing in circumstances where contraception was unsure, non–existent or forbidden and where women needed the legal and religious protection of marriage. They needed this because of their vulnerability in pregnancy and inability to earn wages independently. In this case sexual union had to wait till after marriage. Compatibility between partners had to be tested by means other than cohabitation.

Many people nowadays of course turn this sequence on its head and seek to test out sexual compatibility first and then, reasonably assured of this, go on to discover whether wider meanings are shared. Naturally the traditional method was the better of the two when the risk of pregnancy was high. Now, with contraception and if women can earn more or less as well as men, the latter may be just as logical. However we must not forget that long–lasting sexual compatibility seems to be just as dependent upon *companionship* of ideas as vice–versa (Gorer, 1971). Physical sexual aversion is actually aroused for many people, for instance, when they see their partner's preferred life–style or morals as alien to their own. So it is as well not to pontificate that the modern 'sex first' procedure is universally the better. So long as children are not born of these tentative explorations, there is surely every reason for it to remain a matter of personal preference without moral discrimination either way.

What can be said more dogmatically is that if a person values the quality of his own and his partner's life, then *serious testing* of each other *must* go on before commitment to living together. We are usually capable of knowing so little about ourselves, even less about others, and are so susceptible to rationalization and self–deception, that commitment even with the best of intentions must very frequently contain gross errors of judgement.

Living together has in recent years become so much the mark of the 'free' young person that there sometimes seems to be a social disapproval of or at least pity for, those who do not want to partner in this way. This can mean that people slip into cohabiting with little preparation. However; once living together it is much more difficult to part than when each still has a separate home to go to. Months and even years of life are often wasted by two people who have lost interest in each other but where neither can quite take the plunge to find somewhere else to live and pay the rent for alone.

Another activity that is common today is free wheeling promiscuity sometimes by one, often by both partners (Schofield, 1976). We have noted that premature commitment for life can be mercilessly restrictive and much trial and error exploration must be better than this. But sexual cohabitation plunges partners into such primitive physical feelings that deep mutual trust is essential. A partner's promiscuity always seems to arouse the profoundest pain and mistrust in the other, jealously is too simple a term for the catastrophe of poisoning that is felt. Whether we like it or not this seems to be our fate as human beings. Some 'injured' people seem to stand it better than others but I myself doubt if anyone really regains their loving after being assailed by repeated promiscuity.

Being in Love and Adult Identity

When thinking about commitment in love and its errors of judgement, it is worthwhile to look back at the previous chapter and its discussion of adult *identity*. When a person has found that personal patterning of roles in work and friendship which gives that sense of wholeness called identity, then he is in a good position to measure his losses and gains in these, to set against the values of a passionate love commitment. This may seem to be a calculating way of looking at such romantic matters. But lack of thought is surely silly; if a person's identity of himself is still amorphous or ill formed, then he has no steadying measure. He will be prone to plunge into a relationship with little self–esteem

191

or stable sense of what he is contributing. Under such circumstances he is likely to be caught in a relationship which is based on infatuations with outstanding parts of another person's character which highlight weaknesses felt in his own. In time, this must become a depleting experience. Then, struggling to find his integrity and self–esteem or adult identity in other pursuits, he tears at his love relationship as if it were a prison.

If a couple have committed themselves to lifelong marriage under such circumstances, and even more if there are children, the consequences can be dreadful. This is not just an abstract piece of theorizing. It is known, for instance (Bernard, 1972), that teenage marriages, before adult identity has been clarified, are much more likely to become unsatisfactory and break up than are those contracted during or after the mid–twenties.

Later in the chapter it will be argued that it may be a woman's lack of opportunity to find an adult identity before marriage that makes for the greater preponderance of women unsatisfied by matrimony compared to men. Erikson (1964, 1968) particularly, has stressed the importance of *finding this identity before commitment to marriage*. This time sequence is important, of course, only where the question of long–term commitment is concerned. It does not mean that satisfaction in work and leisure must be found before embarking upon love affairs, and for most people these happen concurrently.

These considerations about identity stress one further point. In order to find one's identity it seems centrally important to be *able to be alone*. Not necessarily alone socially, but *alone with one's conscience*, clinging to no one's opinion in order to carry out one's evaluations. This may sound an invitation to moral arrogance. But to be responsible for one's opinions and decisions does not mean that we need not learn from others, which surely is the mark of arrogance. It is certainly a high demand on ordinary people, and few of us can be truly honest with ourselves in consistent solitude of thought. But without it in large measure we are but mimics and childlike half–adults (Bailey, 1984; Shor & Sanville, 1978; Smelser & Erikson, 1980; Stevens, 1979).

Legal Marriage

Marriage vows are made in the presence of a couple of witnesses in a few minutes, yet it deeply binds *for life* most people, for better or worse. Is such a legally binding oath really valid for human beings today or is it yet another useless relic? It is obvious why marriage has been made so solemn and binding. Not only is it usually born of romantic love and reciprocal appreciation tested out over months and often years, but it is also a *public act*, symbolizing a profound personal psychological event. Each partner is moving further *out of his own family* to become also a member of another, as well as creating a third. Added to this, friends, fellow workers and the state formally recognize the change. All these parties will be involved. Friends will be gained and lost, work loyalties will be changed; and the state with its tax laws views married people differently from mere cohabitees. Thus, quite apart from the partners' personal attachment to each other and the feelings of conscience and responsibility that this entails, the couple are surrounded by enormous *external* expectations that the marriage vow will be kept until death.

Most central of all we must not forget this last point, that the marriage vow as we know it today is *until death us do part*. It is 'for ever' with no ifs and buts or maybe's about a time limit. This is personally very important for it means that each partner is saying to the other, 'You need not worry about me leaving you, I won't'. They are saying, 'You can trust me'. Yet a quarter of marriages, in Britain at least, end in divorce. Something strange must be happening.

The solemn vows are all based on the decisions of two young people who can hardly know what is in store for themselves. They can only have a fleeting, secondhand knowledge of the demands of later life, the turbulence of child–bearing, the boring repetition of home, the burdensome anxiety of wage earning, the yearning for new lives in middle age and the restrictions of infirmity. It is almost strange that it ever works at all.

We are now really addressing a question which is statistical and sociological because marriage is a legally recognized cultural institution shared by many people. It is thus outside our scope. But, as it has such a deep personal bearing on individual development I must be an amateur social historian for a few paragraphs to outline the social setting individuals find themselves in.

Monogomous Marriage, a Brief History

The institution of marriage for life, based only upon the decision of two partners who are romantically in love, is, of course, not universal. There is polygamy, polyandry, patriarchy, endogamy, exogamy, matriarchy, patrilocality, matrilocality, and so on. Each one is considered as quite moral in its own society. What is more, the criterion of romantic love for a marriage decision is of fairly recent origin, even in the West.

In times past, probably because economic survival depended upon the extended family, formal marriage, if it occurred at all, was often more a matter of the convenience of the *two families* than of individual fulfilment, so that leading relatives were the deciding judges. This is not to say that individual romantic love was unrecognized. For instance, the Greeks understood it in their idea of searching for 'the other half', but this could be about homosexual as much as, or more than, heterosexual love. Other considerations seemed more important, however, when contracting a marriage. There are a few tales of romantic love in the Old Testament but these are rare compared to arranged marriages.

The New Testament says little about marriage and romantic love; however their origin in the West can be traced to it. After centuries of turbulence in the Dark Ages, North-west Europe found a new stability and relative peace based on the growth of feudalism in the eleventh and twelfth centuries.

Possibly because of this social softening and stability there arose at the same time a new religious emphasis on Christ's gentle humanity and suffering. From this grew a gratitude to and worship of his mother, the Virgin Mary. Until then she had played only a small part in doctrine. This Mariolatry originated in France, but spread fast and is now enshrined for all to see in the great cathedrals of north–west Europe. It was paralleled by a secular movement in the cult of the troubadours. The knightly ideal that they celebrated was that of a man who would devote his life to his lady, content to carry out her commands and earn her favours. It was for the *lady* to decide whether he was worthy of her. This is a development of historic importance.

But the relationship celebrated by the troubadours was still irrelevant to marriage, even often adulterous. Their romances perhaps provided an escape from the harsh realities of the marital customs of feudal Europe. However, the idea that a woman is no longer a chatel, that she has the power to say, 'No', to the man's advances, is seen here for the first time. This can be seen as a step away from male dominance, and certainly the whole courtly ideal contributes towards our present–day aspirations concerning the relationship between men and women. This relationship was argued throughout Mediaeval literature in the 'Marriage debates', and this is seen perhaps in its gentlest and wisest, as well as bawdy, form in Chaucer's Canterbury Tales. We must, of course, remember that we know directly only of the concerns of literate or wealthy people; there is evidence, for instance, that amongst commoners working women were much more equal partners with their men. At all events, romantic love, originally an ideal, standing in time and space outside the ties of marriage and family, has been *grafted onto the main stem of marriage* and child–rearing. How far this can work is another question.

In Elizabethan and Jacobean writing the idea of mutual love is widely expressed. It is perhaps only then that reciprocal appreciation, of two lovers *equally* concerned for each other, came to be expressed. This is particularly apparent in the poetry of John Donne. But it was several centuries before such ideas could be solidly confirmed in social reality. Seventeenth–century divines played their part in enjoining men to honour their wives, but a primary factor in the change must have been economic. With the Industrial Revolution, as people became mobile in their search for work, the nuclear family tended to replace the extended one as the social unit. When this occurred individual men and women had to find their own partners to start a family, separated perhaps by hundreds of miles from their kin. Furthermore, industrialization meant that *women came to be employed* in large numbers outside their families and being independently employed they could say, 'No', to their suitors without fear of family repercussions or destitution.

Women's Fate in Marriage Compared to Men

Rapid progress towards equal opportunity in employment might have been the happy ending of woman's rise to equality with men in marriage, but it is not quite. How do women fare in marriage compared to men today? The statistical and sociological evidence cogently amassed by Bernard (1972) suggests that women still do badly out of the deal. Married men, for instance, tend to live longer, are physically healthier and less prone to mental disorders than are bachelors of comparable age. Married women with children, on the other hand, tend to have shorter life–expectancy, and suffer very significantly more from mental disorders than do childless women of like age. Thus the worse affected are those who have been *mothers* of families. Married women who have not had families fare better. Bernard's work is predominantly American but is worthy of note in Europe and elsewhere.

Why should married women be disadvantaged like this? Among other things, perhaps this is due to social expectations which give less satisfaction to wives. For instance, a man is expected primarily to pursue his work and his marriage should follow after it. This is a fundamental way in which men differ from their spouses. Married *women* are still symbolically *expected* to *throw away their identity* by changing their names on marriage. They are often not expected to have 'found themselves' before marriage, nor to maintain identity other than as wife and mother after it. 'Wife and mother' is obviously honourable enough, but it does mean that a woman is likely to be financially dependent on her husband for many years. This financial dependence must, for many, encourage fixation upon old infantile dependencies. We know that mental disorders involve the sterile repetition of infantile patterns of function. And it is predominantly mental disorders that Bernard reports as the fate of wives and mothers. Women often wilt into wives (Grosskopf, 1983; Miller, 1976; Oakley, 1974a & b; Oakley & Mitchell, 1978).

These are very serious social questions. It is the women's movement who have clearly drawn attention to this problem in the first place. They are matters of long–term social policy about which I am not equipped to argue.

Remember too that it is often for many men as well as women

that legal marriage entails more shrinkage and burden than joy. Let us look at it in a little more detail.

Marriage as a Life Long Partnership

We have just been arguing that legal marriage as we know it has grown through times when contraception, other than by religious or legal strictures, was virtually unknown or at least very chancy.

Now that contraception can be practiced pretty safely and women have almost equal job opportunities, do we need marriage as it is? (Grosskopf, 1983; Nichols, J. 1975; Oakley and Mitchell, 1978).

We will of course not try to answer the question here, but we can think about it. Considering the *social relations* in a marriage again. A legal ceremony in a church signifies the vows of a couple to each other in relation to the following, God, the State, relatives and friends.

We will not consider the relationship to God here. For some this is very important, for others not at all. Next what about marriage and the State. Is there any real value in the State and its legal system having any more part to play when a couple make a marriage contract, than when two friends make a verbal contract with each other? Certain aspects of the State, like Revenue departments, at least in Britain, are as backward as any male dominated organization in according equal rights to married women. So married status makes a difference here. But beyond this is there any point in a couple being legally married?

The answer for a childless couple may well be; probably not. But it is a very different matter where *children* are involved and where women, if they are to mother happily, probably have to give up remunerative work for a time. They, as well as the children, vitally need legal protection against desertion and neglect.

Thus the essence of legal marriage must really lie in the protection of mothers and children. This necessity will be restressed in our next chapter (Miller, 1976; Oakley, 1974a & b; Oakley & Mitchell, 1978).

Young people nowadays quite often get married just before or just after, starting a family. They may have been partners for a long time and then want to affirm to each other, their families and their friends that they trust each other to stay together to bring up children. Such a method can surely not be condemned except perhaps by those with certain religious scruples.

Even so, there is still an important function in marriage for partners who intend to live together and for one reason or another are not going to have children. It seems also right in this case to say to each other, 'it is for life, you need not worry, you can trust me'. This can be done privately of course, but it is also important to state it publicly to family and friends so that they know what is happening. So far there is no socially recognized event that carries out these functions except by also calling in the church or Registrar to make it a legal marriage which, as we have seen, may otherwise be valueless socially and emotionally.

Perhaps a practice of announcing a partnership to family, friends and colleagues will grow. It could clear some of the uncertainty that surrounds many living partners today who have no appropriate public ceremony in which to make their vows.

Long Term Marriage Harmony and Discord

We have noted, only broadly, how marriage involves a loosening of self–boundaries. Some individuals change, not necessarily for the better, in the formation of a married team. It is thus a crisis of life, not usually felt as such at the time because of the happiness of new–found love. Later, however, extremely painful and violent crises can occur very frequently indeed. It is my, perhaps over gloomy, impression that marriages without long periods of crisis, or without lapsing into sterile deadlock, are perhaps the exception rather than the rule. This seems to contradict more optimistic statistical findings such as those by Gorer (1971). But these are usually based on what people consciously say about their marriages and do not explore the subtle shrinkages of experience that so often seem to occur.

This raises a vital question for most married couples. What is in the ascendant when couples are in harmony, as compared to when marriages break into crisis which may then sink into painful sterility? I will introduce a few general points about this in some brief descriptions of married couples.

Mr and Mrs A

The report says:

I have never seen any show of love on either side, there was no kiss goodbye or welcome in the evenings. They seemed to have nothing in common, and therefore no subject to discuss jointly. His work and hobbies were his only topic of conversation. These bored his wife, and she spent more and more time with her mother, who came to live with them.

There were rows over Mrs A's mother. He resented her interference but, instead of telling his mother–in–law to mind her own business, he told his wife to speak to her. Naturally she didn't or couldn't, and there were more rows. He went his way, she went hers. Eventually the grandmother died and the children left home but Mr and Mrs A stayed the same. They considered living apart, but decided that this would not be possible economically.

Mr and Mrs B

One day Mr B outlined his reasons for seeming so depressed. He talked for several hours saying, 'I can't talk to her and she doesn't seem to understand'. On another occasion, Mrs B gave a long account of hours spent locked in the bathroom crying into a towel. I asked why she needed to hide her feelings and she said, 'I wouldn't let him see me so upset'.

They never used words of endearment and Mrs B said they all ended on the wedding day. Her husband is charming to women. In company he would say to his wife, 'Why can't you look like that?' This might have gratified the other woman but hurt his wife terribly. Finally, Mrs B started an affair with a neighbour; her husband wouldn't believe it even though everyone else knew what was happening. They have separated now.

Mr and Mrs C

Mr C was a lecturer at a college and gave a lot of time to the drama society. His wife felt left out and neglected; being very pretty she soon attracted other men, but he spent more and more time in rehearsals. The climax came when she slapped him on the stage one evening in front of all the students. Her parents encouraged her to leave him and return home to them. She did this and was soon having an affair with another man.

Mr and Mrs X

Mrs X is a driving sort of person and her husband much quieter. But if she wears the trousers it is because Mr X is content to let her do so, and she enjoys it. They have endured considerable hardship together, especially when he was out of work. But this seems to have made them more united. Now in middle age they still show all the outward signs of being in love. They say that they have few major quarrels and most decisions are reached amicably. One suspects that she makes most of these while he acquiesces, but when there is something about which he feels particularly strongly she would not attempt to argue.

The whole family is doted upon by both sets of in–laws. They have helped them out financially in the past, but do not seem to intrude.

Mr and Mrs Y

In many ways husband and wife are very different. Her liveliness can at times be overwhelming. He is quieter and seems to find her liveliness attractive. His quietness seems to counteract her boisterous nature in a way that is comfortable. They often have different opinions and argue, but one senses a tolerance and respect for each other's views.

They say they hardly ever have a row. When they are on the point of one they say they go to each other and swallow their pride until the bad feeling subsides. He is interested in her clothes and ideas; he likes to talk about his work. Both are rather eccentric and vague, but in different ways, so that between them they get things done efficiently. They are both loyal to each other in every sense of the word.

Mr and Mrs Z

Mrs Z is gay, dramatic and untidy. Her husband is critical and can be rude; he is something of a snob. One might not think they would mix well, but they seem to have an inner respect for each other. They appear to me as a couple and not as two individuals trying to keep their own side up.

When he is insulting or flies off the handle, she will stand up for herself and tell him he is wrong if she thinks he is. Rows do not seem to last; perhaps they just have a sense of proportion. Mrs Z says that he upsets her at times. 'Sometimes I could throw him down the drain, I don't know why he says the things he does, but then there must be things about me he hates. I know I wouldn't really change him.'

The cases of Mr and Mrs A, B and C show an underlying *splitting* of feeling occurring. Each tends to see the other in pure good/bad terms – now on the bad side only; each openly hates the other. All appreciation has gone, so they are gripped by the feeling that their partner is just bad to them. They see themselves, on the other hand, as poor sufferers struggling to keep their end up. Faults are reeled off, resentments of long ago are unearthed and chewed over inexorably. Primitive, paranoid mechanisms predominate in the partners feelings about each other. This, however, does not necessarily extend to other people outside the marriage. They do not manifest themselves as generally paranoid characters, they are enmeshed in these feelings only in the intimate partnership. Malign *regression* has occurred in a way that it is confined to the marriage. It is *neurotic, not psychotic* where the paranoia would have spread to more general delusions. We have noted that this can happen to anyone on all sorts of occasions. But various factors conspire in marriage to encourage such tendencies. There is its intimacy, its dependence, the loosening of taboos on eroticism, and the loss of boundaries in becoming a team. It has often been stressed in previous chapters that such regression is vital for full–bloodedness and deep developments in mental functioning. But is also means that there is a proneness to violent splitting, projection, and denial of emotions. These paranoid processes are chronic in such marriages.

It is also noticeable that the *partners' parents* were often intrusive with the A, B and C's but not the X, Y and Z's. For whatever reasons, the former group do not seem to have resolved their *childhood ties* to their parents and have been unable to form autonomous adult identities to take their marriages. Parents often play very active and malignant parts in the splitting processes which occur in marriage. In such cases a person often turns back to his parents to be all 'good' to them, while the partner is rejected as bad.

Turning now to the X, Y and Z's we see that splitting into good and bad has not solidified into a way of life. This does not mean that hatred is never experienced. To re–quote Mrs Z, 'Sometimes I could throw him down the drain . . . but then there must be things about me he hates. I know I wouldn't really change him.' Here anger emerges, but close after it comes *depressive concern* ('There must be things about me he hates'). This is followed by appreciation, the couple simply like each other. Splitting has not taken hold, but rather an integration of feelings and ideas creates perspective and a sense of proportion. This has succinctly been described by Dicks as 'containing hatred in a framework of love' (Dicks, 1967; Bailey, 1984; Skynner, 1976; Thomas and Collard, 1979). In happy marriages, partners are *kind to each other*.

It must remain an open question why benign processes such as these occur in some marriages and not in others. Certainly, individuals bring propensities from their childhood. But just as important must be each person's choice of partner and their intermingling of feelings as they go through life. Those who strike a note of appreciation and protective sympathy for each other, develop an underlying *respect* which is emphasized in Mr and Mrs X, Y and Z. They liked each other. When this key is not maintained couples tend to retire hurt, bitter and disillusioned, and paranoid splitting becomes entrenched. Dicks (1967) particularly stresses the omnipresence of mechanisms like these in marital breakdown.

FURTHER READING

1   Bailey, C. (1984), *A Loving Conspiracy* (London: Quartet). About marriage generally, pleasant reading.
2   Bernard, J. (1972), *The Future of Marriage* (Harmondsworth: Penguin). Written polemically but full of well documented references to research, wisdom presented enjoyably.
3   Dicks, H. V. (1967), *Marital Tensions* (London: Routledge & Kegan Paul). Quite old now but still one of the best books on marital therapy. Its introductions are very good for general interest.
4   Oakley, A. (1974), *Housewife* (Harmondsworth: Penguin). This is a description by one of our best and most forthright authors on women's problems.
5   Tunnadine, P. (1983), *The Making of Love* (London: Jonathan Cape). Enjoyable reading from someone with long experience of tackling marital problems.

CHAPTER 12

Parenthood

Parenthood as a Development Sequence

We have now come full circle. We return to pregnancy and the early years of childhood, but this time we will identify mostly with the parents. Parents still carry their own childhood experiences with them, but are now responsible for bringing up the next generation. It will be useful to browse again through the early chapters of this book to recall what this responsibility entails. With that in mind the thesis of this chapter is: *As a child grows, so it is necessary for a parent to change in sympathetic responsiveness to his/her child. This entails not only being intuitively aware of himself at equivalent ages to the child, but also being aware of his own life and his child's life as separate and with differing present–day problems.*

It seems to be only through being sympathetic yet separate that a parent can assist his child when he cannot act for himself, yet allow him to be free when he can. If a parent fails in this at any stage, he either does not assist him when needed, and the child is distressed, or conversely, he chokes initiative. Both lack of or over much care have detrimental consequences for development. Parents are called upon to develop in parallel with their children. Their job is *to help their children into adulthood.*

This is the point of view implicitly taken by many writers on bringing up children (Kelmer–Pringle, 1974; Winnicott, 1964; Carter & McGoldrick, 1980; Walsh, 1982). It is most explicitly stated by the psychoanalyst Benedek (Anthony and Benedek, 1970). Naturally every parent fails to live up to the criteria of good parenthood to a greater or less degree. They have their own lives to lead and cannot be aware of everything about their child. What is more a child cannot develop independence if scrutiny is overwhelming; *thoughtful neglect* is not only inevitable but also necessary for a child to be unspoilt. But, if childhood is to be

203

deeply enjoyed by the child and the parents, then the parents must take their responsibilities seriously. Quite apart from anything else, it is only when being responsible that any parent will deeply enjoy himself. But it is exhausting and also limits many other activities that invite and press themselves upon people today.

Parents in Society

In societies with little technology high fertility is balanced by high mortality. This was certainly the case in Britain until a century ago, when death was common and a child could hardly expect to live until old age. People then tended to be fatalistic and life lay in the hands of God. Although many parents loved their children much as they do today, they could not be so hopeful for them. Probably partly because of this carelessness was common; there is much evidence that neglect, exploitation and cruelty, especially in Europe, was rife (de Mause, 1974). However, non–technological rural living and extended families meant that children could safely wander and explore the homes of relatives nearby. Responsibilities of care could be spread.

In modern urban communities families are hemmed in by lethal motor traffic if not cooped up in high flats. Parents are under continued stress to be vigilant. What is more, isolation from relatives may give freedom from interference but throws parents on their own resources. They are strained by the knowledge that they have only a few hours in which to relax or be ill, for there may be no one else to help (Oakley, 1974a).

Medical and dietary knowledge, making for increased life expectancy, means that parents can optimistically throw themselves into the enjoyment of bringing up children. But there is a price to pay in heavier guilt if anything goes wrong. The predicament of parenthood is epitomized by birth control itself: there is greater freedom to choose but the parents must then bear anxiety and guilt for the consequences of a birth by themselves.

In our present society it is mothers who bear the brunt of this most directly. As we saw in the last chapter, mothering is a high–risk occupation for mental disturbance and debility. It is very often devalued. When asked who they are, many women do not expect to be honoured or admired but apologize with, 'Oh,

I'm just a mother'. Yet it is largely upon them that the vitality of any society and its future rests (Oakley, 1974 a & b, 1981).

It has been well argued, particularly by the women's movement, that much can be done to free mothers from this lonely burden. Certainly many fathers will have to become more engaged in their families if the burdens of their wives are to be lightened. And other institutions, particularly when considering hours of work, need to change their practices if the isolation of mothers is to end. But our earlier chapters argued that the old nuclear family, maligned as it often is, provides certain vital conditions for the care of children, especially when very young. So that, whatever other rearing means are experimented with, many aspects of this unit are precious to preserve.

What is more, it is only mothers not fathers who are pregnant and lactate; it is women who go deeply and enjoyably into maternal preoccupation. So, if a mother is really to enjoy herself, not to mention the child, she *must* expect to temporarily lay aside other ambitions so that she is free to engross herself in parenthood. This argument, which was also made in early chapters, will be unpopular with many people. But I must say it because it seems very important.

Let us look at mothering more closely.

Fantasy and Reality in Having a Child

Primary maternal preoccupation has just been recalled (Winnicott, 1958). You will remember it being described as a life crisis where old modes of adaptation and defence are loosened so that a woman cannot be as vigilant as she was; she must become more dependent upon others to withdraw and regress. The baby inside her womb is likely to become the focus of vivid bodily feelings, fantasies and imaginings about the future. A mother is thus prone to dream of things for her child that she herself has missed, or to dread that her own bad experiences, or unwanted characteristics will re–emerge in the new life. As the baby is inside and part of her, it becomes part of her own inner world of affect–laden imagery. Put dramatically her baby *is* her unconscious imagining. At the same time her infant is a new creation, an entity of his own who is not just a fantasy in the mind of his mother. These two, fantasy and reality, are in continuous interplay in the mind, not only before birth but long into later life.

The closeness of a baby to his mother's inner world of feeling and fantasy means that her exchange with him is usually warm, deeply emotional and colourful. But it also means that if she is to be healthy for herself and child, she must be continuously, if subconsciously, testing her fantasy out against the baby as he really presents himself before her eyes. For instance, a mother may at a glance catch a glimmer of resemblance between him and her own father. The baby may then be endowed with feelings that rise up in her from ideas about her father. But next the baby turns his head and looks like her mother–in–law and this arouses its own particular imagery. Then the child laughs in his own special way and impresses himself upon his mother as a conscious living person in his own right. The mother's fantasies may well emerge again, but they will now be altered slightly by her realistic perceptions so that a richer representation of her baby slowly integrates itself for her.

Such a close investment between inner and outer world means a delicate balance between the two which is readily thrown into disorder. Thus maternal breakdown in infancy is not uncommon. Sometimes this takes the form of a gross breakdown of the differentiation between her fantasy and external reality as in a *puerperal psychosis*. Of more frequent occurrence are puerperal *depressions*. Here, unlike psychosis, a mother usually experiences a flatness because she cannot endow her child with enough rich feeling and fantasy from her inner world. When this happens a mother can feel dreadfully worthless, quite apart from the distress it may cause her child. Depression of this kind, in transient form at least, seems to be almost inevitable and must be regarded as part of the normal crisis of early mothering. When it becomes chronic skilled help is most advisable. Talking to a Health Visitor is a good way to start.

Some mothers do not break down in their functioning, but unselfconsciously turn their children into the playthings of their fantasy in a chronic way so that it becomes a life–style. Here are a couple of examples:

A woman, whose own mother had become pregnant two months after she herself was born, and then rather neglected her, produced several children and tried to keep them as babies as long as she could. She seemed, through them, to have the gratification of a long babyhood which she had herself missed. She lost interest in them as soon as they began to walk.

Two parents lost their five children in a fire. They then had five more children, who turned out to be of the same sex as the first. The children were given the same names as those who had died. When the fifth was born the whole family with great rejoicing emigrated to a commonwealth country, just as the parents had planned to do just before the disastrous fire of years previously.

Psychiatric literature is full of other instances (Anthony and Benedek, 1970; Bell and Vogel, 1968; Laing and Esterson, 1964; Lidz, 1963). It must be stressed, however, that by the nature of parents' fantasy this is not just a question for psychiatrists. Every child is in some measure the sport of his mother's madness; it is part of the fate of being a child dependent on adults. But where a mother continues to match fantasy with an awareness of her real child then the experience can be rich for both.

Multiple Responsibility in Mothering

The business of mothering is a continuous *to and fro* between mother and child. Her life is acting and reacting from dawn to dusk, so that both she and her child are enjoying each other, yet are self–willed, tired and angry. When a mother loses this responsiveness, then the child is *left alone*. Children are glad of this in small doses, but if prolonged it becomes unbearable. As a child grows older the areas of activity where this 'to and fro' is required become wide and various.

Perhaps the most common steroetyped idea of a mother is that of cuddling and nursing a young baby. This may seem too trivial an oversimplification even to mention. But many young women are so cozily aroused by this idea that they envisage little else. When their children are no longer babies, they can find themselves at a loss and become lethargic, bored and depressed. The children for their part are then left understimulated, drifting about lonely and disillusioned. This pattern is frequently reported by despairing health visitors and social workers. Fathers also may tend to limit their ideas to this fantasy of infancy. Being unable to envisage other responsibilities for a mother they irritably dismiss their wives' work as trivial. Years after their infancy, many young people complain bitterly that their mothers could only see them as babies and not as they really were. It seems to be endemic as a source of family friction.

One mother epitomized the essential quality of mothering as *multiplicity*. If we cast our minds back to the previous chapters on infancy to adolescence, we shall recall how complex the patterning of activities was in those years. A mother is called upon to watch over all these and respond to them as her conscience dictates. This is the responsibility of mothering.

Changing Responsibility with the Growth of Children

As a child grows up, a mother is called upon to develop her responsiveness to match. This entails continuous, new intellectual learning and articulation of feeling. Let us recapitulate some of the salient features of earlier chapters to clarify this (Carter and McGoldrick, 1980).

The first months of life see a mother preoccupied with attuning her body to the physical presence of a new living being. Later in the first year a child is beginning to recognize his mother, and wants to discover his and her boundaries by playing with her. With the second year he is mobile, and a mother finds her will clashing with his. Personal and social discipline begins to become a crucial issue. She must take the initiative. The time has now come for a mother to *help her child on to the next stage of his growth*. This has to go on throughout life.

As childhood progresses the young person begins to talk and explore. When a mother has time and inclination, she can spend hours in friendly investigation and explanation which is deeply satisfying to both. If she has neither, they can both lapse into irritated boredom.

About this time a mother may be having another baby. She now will need to relax into her pregnant preoccupations but at the same time keep in tune with, discipline, provide for and enjoy her first child. After the birth she has the broken nights of feeding and tired days with both children. Then she has their jealousies to negotiate and find ways in which they can be comrades together (Dunn, 1984).

By the age of three or so the child will probably be seeking out other children. This means a mother must broaden her horizons to befriend other mothers and be liked and trusted by their children. For this to happen she has to extend her modes of thought in order to be able to talk to several children at once. The breadth of response must be even wider if a mother has several children of her own of different ages. It is only necessary to be at tea with, say, half a dozen children of different ages to see how exhausting this can be.

Coming out of infancy, the child becomes a sexual person. Parents' sexual feelings towards their children are often dim and little talked about, but most mothers are privately aware of these and discipline themselves to cope. For instance, mothers often find that they have no trouble in cuddling and kissing their children of both sexes alike until they are three years old or so. Some then are slightly shocked to find themselves being a bit disgusted by too many caresses from their daughters while still enjoying kisses from their sons. Both sexes want these caresses, and the daughters get hurt if rebuffed while brothers are favoured. One mother, equally fond of all her children, said she found herself rather guiltily cuddling her son in secret when her daughters were in another room. The quite frequent incidence of sexual child abuse demonstrates how hard it is for some people to keep their incestuous sexual wishes contained.

With school age comes the question of following intellectual progress and finding a working relationship with teachers. Soon after this the child tends to turn more towards the private company of his friends. Then begins the long process of losing intimacy with a child as he proceeds into puberty. In adolescence all a mother's old ways of doing things are likely to be called into question. It becomes a 'day of judgement' for her years of motherhood, so that she may be fraught with guilty worries about her failures.

In so far as a mother has kept in touch with her child, recognizing his and her own independent existence, however much she may muddle along the way, both of them will probably have found life profoundly worth while. The one–sided parent–child relationship may recede, but the friendship, begun years before, in the cuddling and chatter of infancy is likely to remain.

Going to Work

When technology was limited and life short a young woman could expect the rest of her life to be taken up by nursing children and domestic economy, that is if she did not need employment to earn bread. A great deal of intelligence had to be exercised at home. Technology has given women more time to enjoy their children and also many years when mothering is no longer full–time work. It has thus made the apparent burdens of motherhood less obvious. It has already been mentioned that there seems to be a prevailing climate of disrespect for mothering. This is not only amongst men, as has been a tradition for millennia in the male–dominated West, but now also more stridently from women engrossed in their own professional aspirations (Black, 1983).

Under these pressures from outside, as well as from their own wishes to find a worthy identity and a need for money, many mothers are torn, from pregnancy onwards, between their family and work.

We have already noted that it will not be popular with some readers, but the argument of this book must be that if a woman is to fully enjoy being a mother, she needs to reckon to put aside other work completely for the first months of a child's life (unless she can work at home). This perhaps optimally continues until at least social control with her child has been well established, so that he has enough independence to go to nursery school at about three years old. Even then, full–time employment may not be possible unless a father can take part–time work, or another consistent person is available. However, after a child has been at primary school for some time, there is evidence (Davie, Butler and Goldstein, 1972) that children especially in their teens *positively benefit* from both parents having the wider experience and freedom which is provided by other work.

These injunctions are severe and many women, not only those living without husbands who must go out to work, find them impossible to meet. This is especially so for those in the full spate of realizing their professional ambitions. People in such circumstances must surely seriously question the possibilities in front of them. They could perhaps decide not to have children at all. Or they must accept that they will not be their child's only mother in the fullest meaning of the word, for others will have taken over many of a mother's functions. This may not be at all disastrous for a child, there is a lot of evidence that it often is not, but the mothers themselves can feel deeply unfulfilled. It is for each person to decide their compromise; one cannot have one's cake and eat it.

On the other hand, if a woman has steeped herself in rearing children for years of her life, there is no easy solution either. She is left with school–age children who do not need her every attention and are in fact often choked by it. She is likely to be understretched, fundamentally bored and with no identity to ensure her self–esteem. If she stays at home she is prone to wilt into indolence and parasitism. If she seeks work she is likely to be at a disadvantage compared to those of her age who have continued in careers. It is another life crisis which must account for much of the reported mental disturbance in women who become mothers.

The other side of the balance sheet is provided by evidence that second and third careers are becoming more acceptable. Prior experience of being a mother, for instance, is often recognized as invaluable in many jobs, especially in the helping professions. Furthermore, mothers who have devoted themselves to mothering have had the opportunity to enjoy something which no others, men particularly, can possibly have had.

The Father in Modern Society

Just as for women, modern society has begun to transform expectations about fathers in a family. This has been less acute than for women, for a man's work identity can still progress consistently even though he has a family. But machines have diminished the ascendancy of the male's superior muscular strength and woman have proved that, when prejudices die, they can do most jobs demanding intelligence as well as men. The male dominance that has characterized Western culture for all recorded history is cracking. As a consequence, there is an underlying unsureness about male functioning, especially in fathering. This has been but little singled out for special study until very recently (Parke, 1981; Backett, 1982; Cath, 1982; Green, 1978) and is often only included in consideration of the family as a whole. But comments like the following are very common indeed.

I think we men fall over backwards to avoid being authoritarian like our fathers, but when we try to be democratic and discuss things, our wives complain that we are being soft.

My grandfather was certain of himself in his narrow confines. But my father had the whole of his world crash about him when he was a young man in the First World War. I don't think he has ever got over it. The only thing he could talk about with pride was the war. I am a bit luckier, but I do wonder where we go from here.

The problem of fathering is, as we have just mentioned, perhaps reflected in recent psychological literature. The first decades of this century were full of discussions, particularly by Freud, of childhood ideas about fathers and their importance in the growth and pathology of the individual. Forty years ago or so the pendulum swung to interest in mothering. For many years very few writers thought it necessary to consider fathering as a central issue. More recent work on this often stresses the point made here, that there is uncertainty about fatherly identity (e.g. Erikson, 1963).

Technology and modern knowledge have expressed the delusions and tricks of male dominance so that patriarchism is not realistically viable in a modern family. Here the co–operation of the independent intelligence of husband and wife working together as a team is required. But this democracy is not firmly established either in law or in general social expectations, so that men particularly are unsure of themselves.

The Biological Circumstances of a Family in the Early Months of Childhood. A Father's Part
(Backett, 1982; Cath, 1982; Green, 1978; Lamb, 1981; McKee & O'Brien, 1982; Parke, 1981)

Repeating a familiar point, in order to be viable a mother must, at least in the early months of a child's life, have outside financial support. If she has, say, three children spaced two years apart this means that she will probably be fundamentally dependent on others for up to ten years of her life. Either another individual or an institution like social security must finance her, or she must look to others to care for her children while she works. If a woman is dependent upon a social security agency or other institutional structure, neither she nor her children are likely to reap the benefit of the quick and easy communication with an intimate caring person which, as we mentioned, grows in the intimacy of marriage. So, in general, if mothers want to enjoy being full–time mothers, they need working men to provide for them.

It has been argued that it does not need a man to be this helping partner, a woman will do as well. But this means that a child will not have a male to identify with intimately; the importance of this is obvious for little boys and only slightly less so for girls. Furthermore if the two women are lesbian lovers, as they often will be, the child will have the puzzle of their sexual relationship to contend with as well.

Thus during the early child–rearing years, it is still valuable if a man 'stands at the door of the cave' just as has been expected of him through the ages. The period during which his breadwinning functions are crucial has been reduced. But the burden of a father's responsibility, if anything, rests more heavily than in times past, when he could count on extended family support. After hearing many men feeling most deeply about their families, I think that it is useful to conceive of a variant of Winnicott's (1965) idea and talk about *primary paternal preoccupation*. This would not, of course, be rooted in pregnancy and lactation like maternal preoccupation. But, so long as it is not shattered or avoided, it can be deep and long lasting. It has many and various shapes.

A father must look two ways at once, outwards towards the world of work and inwards to the family. If his wife is to relax his work must be, in a sense, primary. When a man will not, or cannot, work, then for financial reasons the family loses its autonomy. Other agencies must be called in to provide this money earning side of fathering. If, on the other hand, a father takes no interest in his family apart from providing for it through work, then at least it can remain an autonomous unit where its members can respect themselves.

213

It is probably because of the primacy of work that many men are ill equipped for family life. Their zest lies in activities outside the family. They are geared to competition, mutual praise and support from their friends in ways that often seem absurd to women. Furthermore, very many of the skills demanded of men concern technology or large groups of people; these are impersonal and lack intimacy. But when they turn into their families quite different qualities are demanded.

A father will have had no formal education about family matters and, in contrast to a woman, he is unlikely to have learnt very much about them from male conversation or literature. For instance, women's magazines are full of articles about personal problems in a way that does not occur in magazines for men. Many forces seem to conspire to take a man's interest away from thinking about intimate family questions. Yet these will be his main concern as a father. Let us now consider them.

Family Environment

The place where a family settles is usually determined predominantly by the nature of a husband's work. Yet his wife and family will be the first to be affected by where they settle and by the housing that can be afforded from his wages. Thus in this, at least, a man will almost inevitably be held or feel responsible for shortcomings.

The predicament of work and housing is seen most poignantly with immigrant families, who move to alien environments because work is available. It is equally difficult for the many men who are unemployed (Fagin & Little, 1984). As often as not, accommodation can only be found which imprisons a wife and stunts the children. The thousands of tower blocks of recent decades are notorious. Even amongst those who can afford to pick and choose, housing is a very frequent source of a wife's misery and resentment, and hence of her husband's shame. But the house or flat itself is only one consideration. The social climate of the neighbourhood and the availability of schools often turn out to be of paramount importance to a mother and her children. Yet these are often difficult to estimate without living in the district first. Perhaps because only a minority of mothers find themselves in an environment where they are deeply contented, many fathers will feel at least some burden of guilt.

Looking After a Mother

When he turns in towards intimate interplay with his family, a father is perhaps least prepared by his previous learning from other men. In the family he is first of all a husband, so that our earlier considerations about being in love and marriage still apply here. But with pregnancy and primary maternal preoccupation he is also called on to increase his vigilance when his wife has slackened hers. Just as she mothers the baby, so he needs to look after her. He is being *maternal*. When a father fails to develop this, his wife is left alone. In an extended family she would probably have female relatives to turn to, but in our society there is likely to be no one to take her husband's place.

A mother needs to be *held* both physically and psychologically by her husband (Winnicott, 1965). This notion of psychological holding like another common term 'acting as a container', is of course just a metaphor which gives no description of the actual mental activity involved. What is really involved is of course largely to do with conversations between a married couple where a husband listens with sympathy to his wife while at the same time being aware of his own point of view. It may be that he then realizes there are actual things he needs to do for her, as well as to act as discussant and arguer. The most frequent instances of this holding will probably not occur in fraught situations, but simply in day–to–day conversations when a mother wants a second opinion for her ideas about the children on routine matters. To a great extent, opinions will be gleaned from other women during the day. But when intimate and crucial questions arise, a husband will be wanted because he is expected to know the nuances of his family situation better than neighbours.

The most usual time for such second–opinion conversations is in the evening after children have gone to bed. Both husband and wife can then relax and off–load their feelings on to each other without hindrance. This is probably the primary reason for the convention of packing children off to bed at the beginning of the evening.

A husband's failure to hold his wife's urges and anxiety is a frequent cause of complaint. For instance, you will often hear such words as, 'Oh, him, no sooner is he home than he's out in the garden', 'When it suits him he'll be as sweet as pie to the children, but help me with them, never', 'He always lets me do what I want but I never know what he thinks about it all', or 'He doesn't know what I have to cope with, he just comes in and criticizes'.

Fathers no doubt react to their wives' anxieties according to their own characteristic modes of defence, which have been built up over the years. Some husbands, for instance, tend to be scornful by nature. If they really respect and care about their wives, they will probably adopt a mode of chiding which can be gentle and helpful because it absorbs worries while being appreciative. But scorn can be used as a defensive means of ridding the self of emotional involvement, at the same time as shattering a wife's self–esteem. Most of you readers will know at least one martinet husband and mouse wife.

There are other ways a husband may fail to hold his wife. A reverse pattern is quite common in the henpecked husband. Here he may be very attentive to his wife, but frightened and in awe of her. Hence, having no independent vitality of ideas and actions of his own, he fails to contain his wife's impulsiveness, which can then turn to anxiety. Such a henpecked husband is perhaps epitomized by the man who said with admiration in his voice, 'My wife and I are of one mind, hers'.

A Father in Direct Relation to his Children

Earlier in this chapter we discussed how a mother is called upon to keep in touch with her children through each stage of development. The same applies to a father, and if he fails in this he himself has lost something of the pleasure of fathering, whatever happens to the children. In early months he is unlikely to be as intimately involved as his wife because he is probably out all day. Likewise his children will not invest so much passion in him, but he is usually deeply wanted for his own particular style of living. As children grow older the relative importance of fathers and mothers evens out so that he may sometimes be more sought after than his wife. But if he has not been in tune with them from

216

the beginning he will inevitably be ill at ease and something of a stranger. He and his children will have missed a very deep pleasure.

Having two parents makes it easier for a child to integrate his ambivalent feelings. Quite early in life a father becomes important as a separate person in his own right. He is of a different gender from his wife, so that boys and girls can find their own sexual identity both by watching father and mother together and by the subtle ways in which each parent separately relates to them. A father is also likely to be a prime representative of the outside world, while still being an intimate family person. Parents vary, but it is very ofen a father who explains and demonstrates the mechanics of the more distant world to children.

Modern urban society presents one particular problem with regard to this learning from fathers. Children rarely see what he actually does at work. In days gone by most children could be with him in the fields or in his workshop. Such an opportunity is now rare. Fathers disappear at 8 a.m., say, to do something incomprehensible until 5 or 6 p.m. A man can easily say he works in a bank, an office, or factory, but verbal reports are usually meaningless to a child. One has only to see a boy working on a farm with his father to recognize that an experience is lost in urban life. A farmer's son of seven or eight can go with his father, watch him and then do things with him, so that at a very early age he is actually doing useful work himself when on holiday or after school.

Children of a father who works in an office can watch and copy him only in his play, tinkering with the car or in the garden. They cannot experience his primary fathering function of working. It has been suggested that this gap in learning about father's work has contributed towards the devaluation of fathering in Western society (Green, 1978; Parke, 1981). What is more, these paternal models, however fully presented by adults, will be near to useless if a child feels personally unrecognized by his father. He will probably be so consumed with hurt anger at being unnoticed that he will be emotionally destroying the models his father presents, however useful they might be in principle. Thus a father needs to attune himself to his children and appreciate them just as a mother does.

Many men assume that they can safely ignore their children in

217

early infancy, but I am convinced they are mistaken. A father who has mingled with his children from babyhood will not only have the pleasure of seeing and contributing to their growth, but will also know them intimately in all their non–verbal idiosyncrasies. Only when he intuitively knows their modes of thought will he be able to be a teacher when the time comes to impart new knowledge. If he has not known them in their early days he will come to them as a relative foreigner, speaking in a way which is unfamiliar and imposing disciplines which seem arbitrary. This bitter experience is often reported.

Just as mothers have difficulties at different stages of development because of the intrusion of their inner fantasies, so also do fathers. The avoidance of infancy just mentioned is one point. Another common instance is reluctance to participate in toilet training. This possibly matters little except that a mother can then rarely feel supported in her endeavours.

Some fathers find discipline easy and provide a useful backstop for a harassed mother. Others find it nerve–racking and impossible to apply consistently. For instance, it is commonly reported that a father has identified himself with his children against their disciplinary mother, so that he has connived in undermining her authority.

Fathers, like mothers, often find themselves troubled by sexual fantasies about their children. They may enjoy cuddling and kissing their children until they are two or three years old. But after this they can shy off such shows of affection, particularly with their sons, often apparently because of fear of homosexual feelings.

These are generalized anxieties, but a father also has his own personal fantasies which attach themselves to his children, just as happens with a mother. When these are flexible and allow room for him to recognize his children as they really are, they give vividness to his children's life. However, when a particular system of fantasy remains entrenched, it can stunt the growth of a child or at least embitter him. Here are a few examples:

A father was brought up in a severe religious faith. After many rebellions he saw the light and became strictly religious himself. However, his ambivalent feelings about it seemed to get attached to his two sons. The elder was a lively and rather naughty boy, so that the father became convinced he had no good in him. The younger son, on the other hand, was seen as a shining light of innocent virtue. The two boys then reacted in character.

The elder espoused unconventional causes and became a militant atheist, taking every opportunity to torment his father, whom he loathed. The younger son was obedient and tied to his parents. He was quite talented as a musician but could only play church music. By his early twenties he had few friends of his own, and had never taken a girl out.

A girl bore a strong resemblance to her mother, who died when she was quite young. The father doted upon her, but seemed fixed upon the image of his dead wife. When the girl had grown into her teens he would still take her on his knee and say, 'Your mother will never be dead while you are alive'. As a woman she was often invaded by feelings of being like her dead mother, and was overcome with bouts of depression.

A father felt himself to be a failure just as his own father had been. He was determined that his son should escape this fate, so saw to it that he was well educated and encouraged him to choose a career that he would throw his heart into.

This itself created few difficulties, but the father also felt a failure with his wife, and encouraged his son to take his place with her, just as he did over careers. Thus he would ask the boy to mediate between him and his wife, and also to care for her in ways in which he had failed. The boy felt proud and triumphant, but also very disturbed. He remained tied to his parents for many years, at least in part because he felt guiltily responsible for both of them, and unable to leave them to their own devices.

These are just a few suggestive examples. Others will readily spring to the reader's mind.

Because his functions are very diverse and lack the consuming physical unions of motherhood, it is rare that a man breaks down mentally; with the complaint that he is unable to father his children. Our society at present seems to make it easier for a father to opt out of his responsibilities towards his children than it is for his wife. This means not only that a great burden tends to be placed on a mother, but also that many serious problems between fathers and children continue unattended into their adulthood.

Mothering Functions in Being a Father

Earlier we noted that men probably get little information about the intimate personal aspects of family life from other men. We have also noticed that these intimacies are called for in the modern family perhaps more than at any time in the past. The young man will have had his first and most important experiences of intimacy from his own mother in infancy and after. He will have learnt about them literally at his mother's knee. For instance, a young father taking his baby out in a pram and chatting to him will have done it all before. He will in all likelihood have seen his mother doing it with him, and then have copied her when he was a toddler by trundling a trolley about with dolls or teddy bears inside. Boys, when very young at least, identify just as strongly with their mothers as do girls. But, for boys, these identifications are with someone of the opposite sex and we have noted before how children are vehemently conscious of their sexual identity.

Thus by the time they get to school age, boys are usually prone to reject the ways of their mother and sisters as 'cissy'. This is enshrined in our society, at least in later education and in the attitudes of men together.

We could summarize this by saying that there is a tendency to reject, or at least feel anxious about, cross–sex identifications. This seems to be particularly the case with boys and men. But it is also noticeable in women when they are shy of doing things they feel are not in accord with their sex. Some women refuse to attempt to drive a car or mend a light fuse saying it is 'too masculine'. I am convinced that such prejudices in both men and women serve to do nothing but stultify their own and others lives. Bernard (1972) stresses the same point. Such prejudiced people fail to discriminate between their sex, with its particular male or female harmonious functioning, and mere symbols of 'masculinity' or 'femininity'.

When a man comes to marriage and then fatherhood, many of the functions he learnt with and from his mother are called upon to be exercised to the full. In many ways it might be an advantage that he has had little education from other men about such matters in his youth. He can be free of indoctrination to make his own discoveries and find his own best modes of caring and intimacy. But it also means that he may have little support from other men, so that he is lonely, ashamed and unsure when faced with personal or family problems.

What is more, cross–sex identification creates its own problems. A father can turn into something of an 'old woman'. He can fuss and care too exclusively, so that his wife is deprived of his male romanticism towards her. She is then likely to become depressed or disgruntled because her sexual femininity is left unappreciated.

Mothers and Fathers Together

Most of this chapter has been taken up with consideration of parents individually relating to their children. But it has been noted earlier, particularly in discussion of the Oedipus complex, that children need to integrate their experiences of their two parents into a harmony in their minds. A child who experienced violent splits between his parents can feel very disturbed indeed. Not only does he fear for his home's security, but his mental representations of his parents will not marry or function together; they contradict, so that it is hard for him to use them.

Some parents seem to be so sensitive to this that they vow never to disagree in front of the children. It is probably a good rule not to indulge thoughtlessly in rows in front of children because they are not only frightening but also confusing. But when the appearance of agreement is carried to the length of falsehood this creates its own confusion. For children are very sensitive to moods, weaving solitary fantasies about them if nothing is said. Open expression of disagreements, recognizing the other's point of view and phrased in terms that a child can understand, is more likely to calm a child out of his confusion than is forced appearance of solidarity.

Even more striking, in my experience, has been the neglect and hurt imposed on children by some parents who have continued indulging in a 'chronic love affair' with each other. By being only interested in utter devotion to each other, they often seem to set up a barrier of indifference to their children, who grow up hurt, bitter and with a deep feeling that they as people, are a nuisance. Romantic love may be a pre–requisite in providing the impetus and stability to rear children today. But great lovers probably do not make good parents. Romantic love is necessary but not sufficient.

Parents Separating Out

Throughout marriage, with stresses at work, changes in home, bringing up children and getting older, both parents will have been torn in many different directions. Each will have been through many, at least minor, crises by the time the children are well into school age. By then a mother will have to find a new work identity for herself or, in all likelihood, wilt. A father on the other hand, with his work to the forefront throughout, will perhaps be differently stressed but have to change in his own particular way. Under such circumstances, it would seem mere chance if both still find easy harmony together. However devoted and kind a couple may be to each other it seems inevitable that, in certain ways at least, they must separate out to find their own identities.

It has been said, 'Everyone should marry three times, preferably to the same person!' A rather glaring statement but it pithily summarizes the predicament of many people. Let us reflect for a moment upon the value for parents in sharing together the growing up and away of their children. We can also recognize the deep importance to children who know they have parents who are happy together even without their offspring, and remain available to come home to. Such conditions gives precious freedom. With this in mind surely few will argue against the ideal that marriage for life is still worth striving for.

Many couples who have lost intimacy with each other find invaluable help in marital counselling and psychotherapy (Dicks, 1967; Skynner, 1976; Skynner & Cleese, 1983). However there are many couples who do find life more fruitful and less stressful after divorce. And there are many others who can be seen to plod on, often torturing each other, in sad holy deadlock, who might have been better off apart. So we must face the question of divorce. I must here warn the reader that, since I have myself been through a divorce, my ideas may still be distorted even though it occurred long enough ago for many old belligerences to have melted away.

Divorce

We recognized in the last chapter on marriage that it is not surprising that many people, basing their choice of partner upon romantic love, find themselves later to be fundamentally incompatible for intimate life together. With all the will–power in the world human beings are limited in how much they can attune to each other's chemistry (Thomas & Collard, 1979; Weiss, 1975).

Having said this, it would be lazy and dishonest to portray divorce as simply a matter of incompatibility. Every couple, those who stay together as much as those who part, have to manage incompatibilities about each other's characters.

It might be better to say that divorce is due not to incompatibility but to *intolerance* by one partner or the other or both. At least one partner comes to a point of not tolerating the idiosyncracies of the other. It only needs one partner to be intolerant of the marriage for a divorce to take place. They may then leave the married home or drive their partner out and even appear the innocent party. It must also be recognized that the intolerance can have arisen for very good reasons, the spouse may be violent, promiscuous, cruel, dishonest or impossibly perverse. However a partner's intolerance may itself be driven by baser motives, by greed, cruelty, promiscuity or other forms of selfishness.

Peaceful, amicably agreed, divorces may take place, but I myself have never heard of one. They always seem to involve the violent splitting and paranoid hatred that we saw in the unhappy marriages of the last chapter. However hard a couple may try to be considerate, and some succeed well enough especially where their children are concerned, there is always a nasty core at least in the early stages. In this light the idea of simple incompatibility can be seen as a bland euphemism.

Children and Divorce
(Inglis, 1982; Roland, 1973; Tessman, 1978; Weiss, 1975)

Divorce between a childless couple may inevitably be hurtful but the partners are then often freed to find happier lives with few long term ill effects. Where children are concerned, especially when young, the issues are much more serious.

Some partners become so engrossed in justifying themselves that they malignly charm their children to take sides against their spouse. When one realizes that this may be against a parent that a child has loved and needed it can be seen how distorting of his profoundest feelings this can be. Other parents can be so taken up with their battle that they ignore the lonely plight of their children caught in the middle. It is to combat this sort of neglect that many divorce courts must strive everyday.

But very many parents really do continue to care for their children and also to be fair with them about the other partner. This undoubtedly mitigates the catastrophe for many children, especially if they are older.

Let us look briefly at some of the issues that a child must contend with on divorce. Starting with the end result that needs to be prepared for. Any child when coming to adulthood needs to find his integrity as a person separate from his parents. To do this he must be able to *use representations of all his family within himself.* Thus an adult's own parents may be long dead but his memory of them continues alive in his identifications with them. These parental (and also sibling) identifications will inevitably involve conflicts and contradictions between them however happy the family. Those of a person whose parents were divorced are likely to be more stressed in their contradictions. But if the parents themselves acted largely with concern and integrity the task of bringing these identifications together within the mind need not be impossible to negotiate.

In summary: any child, as he grows up, needs to coalesce all his family together in his self. How can this be facilitated even when parents separate?

First let us look at the bad side. When a divorce occurs one parent must leave the children so that he (or she) and they must be depleted by the absence of each other. The other parent is usually then overburdened with responsibilities and hence, often deprived of other opportunities to enrich his (or her) life, can shrink into martyrdom. What is more, divorce is often more apparent than real for, whether they like it or not, parents are still tied. Even if they have opted out of everything else, they are tied by guilt, for it is they who brought their children into the world. Divorce may be necessary and ultimately benefit the children, but they must suffer in their opportunity to identify with their parents.

Most optimal alternatives involve finding ways for each parent to gain his own freedom in ways that are compatible with both parents having continuing, enjoyable contact with the children. This occurs in an environment where the children *can be at home with either parent and with friends of their own.* Children find contact in a strange environment very lonely. Some financially lucky parents manage this by having two homes close by. Although there are many different combinations that can be considered, I have never known any that were not painful, at least in the early stages. Very often mother or father, or both, partner up with someone else so that their child and a step–parent must attune to each other. This is rarely easy, but on the positive side it must be said that, for all the difficulties, this can mean another set of near–relatives which in the long run may be enriching for everyone (Visher, 1979; Wald, 1981).

Whatever new patterns are found it is surely better to work things out with care than to act on impulse ignoring vital issues however 'natural' or 'honest' it may seem.

When parents are trying to work out what to do and are worrying about the effect on their children, they might find this very oversimplified list of priorities for all children useful.

(1) A child needs to be happy with his mother and see that she is enjoying herself.
(2) He needs to be happy with his father and see that he is enjoying himself.
(3) He needs to be happy with his friends in an environment he knows.
(4) He needs to see his parents enjoy themselves together.

Optimally all these require very frequent contact with both mother and father *in a familiar environment* until adolescence at least.

The fourth requirement, parents enjoying each other, must, of course, be more or less intractably ruptured with divorce. It is my impression however that *if the other three requirements are well met*, then it is sad for a child but need not be catastrophic. What is more, parents can still enjoy themselves together *about their children* even if divorced. Our list of priorities points up that consideration of divorce as a simple causal entity is too crude when children are concerned. Rather one has to evaluate and mitigate the many ruptures that may occur. The list also suggests that under some circumstances it might be much better for a child to enjoy being with each of his parents divorced as separate contented people, than to be burdened with two stressed people who keep together unhappily for want of working out alternatives. Fundamentally it must be the responsibility of each parent to decide the way that seems best for themselves, their spouse and their children.

Single Parent Families
(Weiss, 1979)

The past pages emphasizing the values of a nuclear family must, without further consideration, seem unfair on the many single parents who manage successfully to bring up happy children.

Before we look into some of the factors that help to make this possible one particular eventuality must be stressed. It is one thing to set about making a good job of bringing up a family alone because of death or desertion, it is quite another to *actively decide* on having a child without any known father. A woman who consciously and intentionally decides to get pregnant in these circumstances must be operating under a system of fantasies that can only militate against good mothering. We have gone over, in the chapters of this book, the vast multiplicity of a child's needs for continuity of loving care from a few devoted people. From these we can see that only women with, frankly, inordinate grandiosity and conceit could consider mothering alone with equanimity. Such equanimity would only arise from a capacity to *deny* the reality of what is involved to a degree which grossly obliterates realistic concern for a child. This denial of reality must be so wide that it often must come close to psychosis or delinquency.

These are perhaps the harshest words in the book. But they do need consideration because many well meaning people confuse the very real need to strive for help for hard pressed single parents with the giving of encouragement to single women who want to get pregnant and mother for the purposes of crudely indulging their own fantasies.

Having said this let us turn our attention to people who find themselves, because of death or separation, as single parents of either sex. Here are a few thoughts gleaned from people who have expressed it, they are not exhaustive but may serve to stimulate further ideas. When a person finds themself alone with young children they are often so busy that they keep going only by staying in a 'high' slightly manic mood. In time this can crack into utter exhaustion and often illness, which is all the worse because there may be little help for the children. Naturally one must keep going but a parent's prime task must also be to look after themselves. In time the numbness of loss will wear off and it is perhaps then an important part of self–care to *let oneself feel*, to feel deeply, angry, sad, alone, glad, whatever comes. We will say more about mourning in a later chapter. Marital loss is death of a sort and thus an essential occasion for letting oneself mourn. It may be that no one has died but the loss of partnership is just as real. People who are left alone may be glad of the relief from a useless or cruel partner but the aloneness is none the less a loss. *Depression is inevitable*, it is surely best to recognize this.

A single parent will probably feel guilty, even remorseful, for their children growing up without the other parent. One cannot get rid of such feelings but perhaps it is best to strive to keep such guilt in check. For guilt can easily sap a parent's clarity of resolve and they can best serve their children by being forthright and firm.

Without intimacy with a partner, a parent, almost more than anything else, needs friends for cross reference, 'holding' and active help. This can only be done by getting out of the house and positively, quite self consciously, making friends, often with people in a similar situation who also need help so that the friendship can be truly reciprocal. With lots of good friends the burden of parenting can become a joy again. Many people have stressed that *getting into the outside world* is essential to sanity, far more important than hours of solitary devotion to children and a tidy house. Children can probably bear a rather scatty parent who has fun in them better than a solitary depressed one who raises a multitude of anxieties.

This need to be in the world can also be considered when thinking about going out to work. When children are very young the extra money that work brings is probably not worth the exhaustion and discontinuity that is entailed. But very soon, when a child is old enough to enjoy a nursery school say, it is perhaps more important for a mother to have the confidence and refreshment of work than to remain solitarily devoted to her children.

Lastly let us not forget that many single parents may have their own family; also, what is more, their partners may not be totally unavailable unless they have died. Children need relatives and particularly the other parent to belong to and identify with. The other parent may be a long way off but be glad to have their children for long holidays. This can be a joy for the children and free the single parent for a change.

Any single parent who reads these pages will have many other thoughts to add to this brief list, but I hope it will serve as a beginning.

FURTHER READING

1 Carter, E. A. and McGoldrick, M. (1980), *The Family Life Cycle* (New York: Gardiner Press). A good theoretical overview.
2 Dunn, J. (1984), *Sisters and Brothers* (London: Fontana). Likewise brief and simple about siblings growing up together.
3 Fraiberg, S. H. (1977), *Every Child's Birthright* (New York: Basic Books).
4 Oakley, A. (1974), *Housewife* (Harmondsworth: Penguin). Now a near classic about women's predicaments in the home.
5 Oakley, A. (1979), *Becoming a Mother* (Oxford: Martin Robertson).
6 Parke, R. D. (1981), *Fathering* (London: Fontana). Brief and straightforward.

CHAPTER 13

Being Alone

Maturity and the Capacity to be Alone

Bernard's observation that married women, mothers in particular, suffer disproportionate disturbances compared to their husbands, has already suggested to us that finding, keeping and being accepted for one's independent identity is crucial to well–being. Her argument also suggests that it is possibly even more important than ongoing sexual intimacy (Bernard, 1972).

Extending this beyond marriage, we have already discussed briefly that the *capacity to be alone* is very important for growth into maturity. Maturity seems to involve not only responsibility, but also uncomplaining and dispassionate reflectiveness about issues where one is nevertheless emotionally involved: one can then look at a problem from many different points of view and bring them together. In other words, the capacity to be alone is necessary for wisdom. Looking back over the course of life from birth, we can detect a continuous ebb and flow between intimate intermingling with other people on the one hand, and self–contained, even solitary, thought and action on the other. A baby, in particular, usually spends hours alone, looking, listening, sorting things out and trying new ideas. Without such solitary exploration a person can be little more than a cypher, an echo of others with no ideas of his own to contribute. In continuing chronically alone, however, a person must remain withdrawn, creating limited ideas about his world. These stay untested and lack the richness and depth gained from intercourse with others. It seems that growth can only come through an interplay between sociability and solitude.

Winnicott (1965) noted this when he stressed the importance of a mother and child being alone together in the house going about their separate businesses. But apart from this, there seems to have been remarkably little interest either by psychoanalysts or more academic psychiatrists and psychologists on the value of solitude. This is probably because it cannot, of course, be observed in a consulting room, nor is it easy to carry out experiments about it. Likewise the social sciences, being concerned with socialization, have not been very interested either.

Yet is is common knowledge that really original scientific and artistic work is generated largely in solitude. Moral and religious thinkers, too, have known of its necessity for thousands of years. Moses went up Mount Sinai to commune with God alone before bringing down the ten commandments. The Buddha meditated alone under the Bodhi Tree until his enlightenment, and Jesus went into the Wilderness before resolving upon his mission. So, both in the West and even more strongly in the East, there is a long tradition of meditation. Perhaps because of this moral tradition, aloneness has been largely neglected in the theories of scientifically minded psychologists. Those with a more existential approach to life have not neglected it in the same way, perhaps because of existentialism's roots in theology (Tillich, 1952).

The importance of self–appraisal in maturation has shown itself time and again throughout this book. A person needs to copy, listen to and learn from others, as well as experience his effects on others. But, when each new crisis arises, a person is called upon to find a new integrity of his thoughts and roles with others. He must not only loosen his habits and go a bit mad, but also turn inwards to dispassionate observation of himself in his multiplicity. Only then can he evaluate the variety of his ideas and form resolves which lead to a new identity. It is particularly important in youth when a person, at his best, is challenged to find his own personal conscience. But it is also necessary in each crisis throughout later life.

Such developmental changes seem to involve acts of new freedom. With loosened habits one throws off old identifications based, as we recognized in early chapters, on synchrony with and mimicry of others. This done, the next step is to be able to move freely from one point of view to another, reflecting on them. This means remembering one view from the point of view of the next, a new integration of ideas and feeling is taking place. This has to be done largely alone; being too attuned to others makes us mimic them, our freedom of movement within ourselves would then be restricted.

The ability to retreat into the self does not, of course, require the life of a hermit. Engrossment in solitude is lop–sided, a whole life–style is then devoted to one side only of the necessary to and fro between solitude and sociability. What really seems to be important is the ability to maintain one's solitary mode of thought in the presence of others, and to recall others' points of view when alone. Some people may find it helpful to go off alone for long periods of time to think, others can make do by maintaining their own way of thinking in the ordinary course of social life. The capacity to be alone is important, whether in the presence of others or not.

Aloneness is frightening to a great many people, for it easily moves from enjoyable freedom into painful loneliness. It seems that it is fear of loneliness that makes many people avoid like the plague the very solitude that is necessary for growth into integrity and its peace of mind.

Why should loneliness be so frightening? A lonely person cannot be assured of physical comfort or have bodily contact; this is enough to frighten some people. But more than this, he usually has few means of distraction from the insistence of his own imaginings. Fantasy wells up with nothing but the resources of his own mind to check it. No other person is present, so that those prone to find comfort by projection of their unwanted ideas into other people have nowhere to put their feelings, they become empty, lost, and frightened (M. Klein, 1952). What is more there is no checking of ideas by others. Then we are like Robinson Crusoe with only the hills to echo back our own cries, or like a lonely old person with only things in the room to talk to. Not for nothing is solitary confinement deemed one of the worst of punishments. But also like Crusoe, it is only through being alone, looking inwards and outwards, that a person can be in a position to know what thoughts he contains, and hence move through to self–containment and the dignity of knowing that he can look after himself in a physical and also mental way.

Many people seem to avoid loneliness by taking to promiscuous sexuality; they perhaps use eroticism as an anti–depressant. Others do it by drinking, smoking or taking other drugs, by overworking, being a convivial bore or turning on the T.V. all the time. Yet others avoid it by jumping from childhood dependency into marriage. Their integrity is likely to be depleted by this.

Certainly it is my impression that many people do in fact shrink into petty narrow–mindedness after years of marriage. This applies to men, I am sure, as much as women. Women may react with more overt signs of distress to the depletion, whereas men can perhaps more easily slip into complacence. It could be said that some married couples become spoilt as characters when each is not left to find their own way enough.

Apart from the chronic painful quarrelling and unkindness which must be the worst that marriage has to offer, this dullness highlights its most negative aspect. It may, perhaps give some comfort to those who have never married. Let us consider this by a brief comparison of the two states.

Staying Single

A person who has not married may not only never have known consistent adult sexual companionship, but has probably never been through the labour (to use Arendt's term) of producing and rearing young. This is fundamentally biological and shared, in their way, by the lowest of animals. Most of those who have been through it will testify that it is very deeply satisfying indeed. However, if a person is really to enjoy rearing children, many other pursuits have to be limited; there is just neither the time nor money available for them all. A single person can have more leisure to devote to wider social and cultural concerns.

There is still something of an unspoken prejudice amongst the married that relegates single people to the position of failed, second–class citizens. It is as if many people vaguely feel that single people have not only missed something vital in life, which they often have, but that there must also be something wrong or immature about them. This is surely not necessarily the case. Bernard, for instance, makes the point that it is often dependent and immature women, not the self–reliant and thoughtful ones who seem attractive to men because they present little threat to their egocentrisms.

What is more, when it comes to the general exercise and enjoyment of responsibility for human lives, it is an open question whether a single person must fall behind the married. For instance, a school teacher who has helped hundreds of children into adulthood has probably made more contribution to future lives than a parent who has had and reared one or two. Many deep enjoyments of parenthood can be experienced just as much by a spinster or bachelor as by mothers and fathers. A few centuries ago, particularly when the Catholic Church held more sway than it does now, the celibate state was held in higher esteeem than the married. Then, in Europe at least, the pendulum swung the other way. Perhaps this social attitude is now beginning to melt, so that people will feel freer to choose the life that seems most fruitful to them without the fear of scorn if they wish to remain single.

As things stand, however, perhaps in the face of social attitudes that devalue them, some single people do seem to shrink into narrow mindedness as they grow older. I have no formal statistics to support this contention, but the single person's shrinkage often seems different from that of complacent married people. Asexual bachelors easily seem to become prey to fussy impatience and misogyny. They seem to identify their whole selves with an anti–sexual part of their consciences. Others, of a more philander-ing nature, seem often to protest too strongly about the pleasures of not being tied down in life. They seem to be seeking solace from loneliness and depression in sexual athleticism. Middle–aged spinsters are, of course, the subject of many myths. They are suspected of prudery and overvaluing senseless correctness. From this position, they can be subtly destructive in denigrating both men and mothers who do not meet up to their arbitrary rules of decency. My impression is that this is not just a myth; many spinsters do feel sexually and parentally very frustrated. Some seem to get over this disappointment by feeling superior and subtly denigrating other more libidinous people. However, some unmarried people openly recognize their loss and are, I think, the richer for it. These are people who command respect, for having learnt to stand their loneliness and to be self–sufficient; they have a dignity which is very fully human (Langer, 1957; Tanner, 1973).

They can, perhaps, also provide a model to married people. This is because, for a marriage to work over the years, each partner must proceed through several, if not many, life crises. These, as we have seen, must in essence be personal, private and hence solitary, however close a person is to his partner. If a marriage prevents the progress of these lonely crises it is doomed either to break or sink into emotional sterility. Happy marriages seem to rest on each partner maintaining his own integrity alone (as some spinsters or bachelors can) while allowing it also for his partner. Only then is sharing what is fruitful together a freely enjoyable experience.

Now, very briefly, I would like to turn to a subject that involves being alone but is not just about solitude.

Therapeutic Conversations

This chapter has stressed the value for growth of the capacity to be alone. Periods of aloneness seem necessary for integration, especially in times of crisis. However, it has also been noted that withdrawal into the life of a hermit is hardly a solution; it seems to act as a defensive posture against anxiety about relating to other people. Yet compulsive sociability can also be a posture to avoid loneliness. If the arguments throughout the book have any validity, sterile defensive postures seem to be a bane of life. But we cannot rest content even with this conclusion. We have seen in previous chapters that defensive avoidance of anxiety, confusion and pain often seems to be necessary for conscious equanimity. However, it can easily become a self–gratifying and sterile habit which impedes development. For growth to take place, many old defensive postures have to be broken. In some crises this can, no doubt, be achieved largely by a person himself. However, because the postures are largely unconscious the intervention of another person, who is involved but dispassionate, is often necessary to break them.

A conversation which leads to the breaking of a sterile, defensive mental posture can be termed a therapeutic conversation. These can occur informally at any time between parents and children, husbands and wives, and between friends, colleagues or lovers. What is necessary is that two or more people are sincere and trust each other. One person at least must also be intuitively in tune with the other's ways of thinking and be aware of the issues involved. These are times of 'talking straight'.

Informal conversations such as these have, I am sure, been used throughout history in the resolution of critical personal problems. They provide the moments of *creative insight*.

In such moments something like the following takes place. As one talks over a problem one is also thinking and a whole array of *particular* memories and specific ideas come into the minds eye. At the same time these are gathered together and an *abstract* idea emerges which is distilled out of the particulars. The 'new' idea is not only a *condensation* of the old particulars but it is also a *new position* from which to see things for the future. By abstraction one is *wiser*. Such experiences and conversations happen every day when learning or dealing with things, be they mechanical, chemical or interpersonal. When interpersonal we can call them therapeutic conversations. In this case they usually arise out of *gossip* with kindly people. Gossip incidently is the most maligned of activities, when carried out with generosity it is essentially informative about the real world. It is as valuable objectively as any scientific communication, but it is about people not things.

However, a person's defences may be so chronic that they escape the capabilities of ordinary sensitive conversation. For these eventualities, the disciplines of casework, counselling and the psychotherapies have been recently formed into bodies of professional expertise.

These professions have numerous and often conflicting view-points. But I would suggest that all, with varying degrees of success, aim towards the breaking of defensive postures and then usually attempt to provide a setting for a person to test out new patterns of thought and action in sympathetic company.

The behaviour therapist, for instance, takes trouble to diagnose a person's pathological habits (defensive postures) and then carries out retraining procedures to break these and establish more fruitful ones. There is much evidence to show that this works in many instances.

The psychoanalytic therapist, like myself, has a rather different approach. Here a person's spontaneous mental activity is regarded as of central importance. A person is thus left as free as possible, within limits set to prevent disturbances to other people, to say and feel what rises up in his mind. By listening to these thoughts the therapist hopes slowly to become aware of the person's sterile, repetitive postures of thought and behaviour which have become so entrenched as to be aspects of his *character*. He usually notices that these seem to include *repetitions from childhood* of patterns which have continued on into adult life, distorting his perception of himself and of the outside world. These repetitions are technically referred to as *transferences*. They are likely to affect much of waking life, but can also intimately be experienced in the actual therapeutic relationship. The therapist attempts to initiate a break in these repetitions by calling attention to their occurrence. Because the repetitions are usually encrusted by long usage, a person often finds these observations unpleasant and may defensively reject them as ridiculous. But even so, an opportunity has been provided whereby the fears that gave rise to the repeated defensive postures can be seen for what they are, often many years old and now usually invalid. This is an *insight therapy*. It is a long–term, systematic and disciplined extension of the informal therapeutic conversations which have probably been used throughout history.

Therapists of other disciplines argue that it is time–consuming, possible only for a favoured few and not necessarily effective in its results. Here I must rest content to agree that these arguments are partly true. There are many forms of therapy nowadays (very often stemming from psychoanalytic ideas) which, though accused of being rough and ready, often seem to be effective in helping people to shake themselves out of old encrusted postures. However psychoanalysis is the most intimately thorough invest-igation of oneself of any of the therapies and, through the many forms in which it is applied, is the most widely established. What is more, its practice has accumulated a pool of knowledge about the internal conditions of human suffering and happiness which have not been known before in the history of mankind. For this, it is worthy of care and respect.

FURTHER READING

1 Brown, D. and Pedder, J. (1979), *Introduction to Psychotherapy* (London: Tavistock). As it says, a straightforward introduction to the ideas and modes of therapeutic conversations.
2 Casement, P. (1985), *On Learning from the Patient* (London: Tavistock). A more specialized but equally interesting consideration of therapeutic experiences.

CHAPTER 14

Mid–life

Introduction

There is a noticeable sparsity of literature about the sequences of individual development in the adult years of life. Much has been written about childhood and adolescence, some about marriage and parenthood, but not a great deal else until old age. There is voluminous writing of general adult psychology about work, sex and other specific activities through social attitudes and behaviour to studies of stress, and then vast libraries about psychopathology and psychiatry. But these cut sections across time and are not concerned with the movement of internal events as a person goes through life. Perhaps, this sparsity is due to the fact that in childhood it is possible to detect processes of development that are more or less common to all children of a certain age. However, in the adult years people change more idiosyncratically. There may be general sequences of development, as in childhood, but they must be spread out over many years as different people arrive at a stage at very different ages. Thus general patterns are hard to detect. But this should not allow us to ignore the fact that each individual must go through highly personal, complex developments in the whole process of his life. We have seen something of this in parenthood; it must be true in other areas, but we know little about it yet.

However, quite a lot is known about one aspect of development in later years: this centres upon a person's changing awareness of the *time he has got to live*. Let us consider this.

The Individual's Awareness of his Life Span

From infancy until death, each person has something of a sense of his own age relative to others around him. This has far–reaching

237

ramifications. Our early chapters stressed the child's pervasive sense of smallness and incapacity. Later we noted the adolescent's sense of being at a threshold. In his twenties a person establishes himself as a contributor, but is also a learner and servant. In his thirties he usually wishes to be recognized for his contributions, and is less ready to relate to others as a humble learner. He is also usually in the middle of being a parent. Throughout these years he may frequently look back, but is still predominantly looking forward to the future. Then, somewhere about the age of forty, he often begins to realize that life is half over for him. When a person looks equally back and forward, he is aware of middle age. We can detect various threads interweaving in his mind to give him his own particular experience of age. Few of these are very pleasant in themselves. There is the awareness of a younger generation who are thrusting forward wanting to replace him and will be alive when he is dead. There is also recognition of physical ageing in his body organs. This deterioration is the basis of mid–life, as much as growth was in the earlier years. We shall not enter into details of its physiology, but it will be repeatedly mentioned in the following chapters (Bromley, 1974; Hareven & Adams, 1982). Physical deterioration takes us to death. This sets the end–point of our existence. Hence *ideas about death set a boundary to the conception of oneself, and are central to the individual's sense of identity*. Such ideas are critical in mid–life, but must have much earlier antecedents if a person is to mature at all. Let us return to childhood for a moment and trace these (Bromley, 1974; Hareven & Adams, 1982; Norman & Scaramella, 1980).

Development of the Idea of Mortality

It was noted in earlier chapters that ideas about death arise very early in childhood. Certainly children of three and four openly express concern about dead things. Usually with puzzlement, awe and worry they will investigate and ask questions about dead animals and birds. They will similarly ask with anxiety about the death of relatives. The sequences of the discovery of death have been well documented, particularly by Anthony (1971).

The worried emotions children show suggest that they have some dim idea about the implications of death. It cannot be said that they understand what it is like to be dead, for no one knows this for certain. Yet a child feels deeply about dead things. What moves him to this? Remembering our earlier discussions about childhood depression, perhaps something of the following occurs.

A person, a child particularly, sees a dead thing and notices that something he expects to be active is not so. It is non–responsive in every way to its environment and within itself. The child's perception of an *external* dead thing perhaps echoes *inner experiences* of unresponsiveness. In this case death is understood by the individual's own inner feelings. Such inner experiences are frequent, as when one 'feels dead inside', 'cut–off', or ways of relating 'die on us'. We have mentioned the 'death' of relationships at separation on many occasions from infancy onward, and noted that they are usually accompanied by depressive feelings. These small deaths are painful but a natural part of living and give it depth and richness.

However the small deaths that are continually taking place *physically* inside the body itself have not been mentioned. These occur acutely in illnesses of every kind which every child experiences. They also occur in the general wastages of the body cells. Thus every child has dim awareness of breaks in responsiveness, if not from splitting and repressed feelings about broken relationships then from the illnesses of and injury to his own body. All of these present threats to his well being over which he has no control. Anxieties about all of these seem to focus themselves into dread and awe when a dead thing is perceived.

It might be argued that a child's dread is simply a social imitation of adults, and certainly they must reinforce it. But it is so spontaneous in children and arises so often when adults feel little anxiety themselves, that this seems an insufficient explanation.

Children's *conscious* personal worries about death are usually expressed in terms of fears about losing their parents or grandparents. For example:

People die at seventy, you are thirty five mummy. So you won't die for seventy take away thirty five years, what's that? Thirty–five years, oh, I'll be forty one then. Granny is sixty five. Well, she won't die yet. (six–year–old girl).

Children have inevitably experienced fears of being left by their parents, so the most important idea of a death is the loss of these much–needed people.

The child's own death must usually seem such a long way off that it is most frequently disregarded or *denied*. They can also project it, as is manifest sometimes in triumphant glee over dead things. However, quite young children do think about it. For example:

> I don't think I want to be a soldier when I grow up, because I might get killed (four–year–old boy).

Children who have nearly died are often preoccupied with worrying about their own mortality when still quite young. For example:

> I was seven when I was knocked down by a car and was unconscious for a week. I remember being preoccupied with ideas of my death through my school days. I had a pervasive feeling of living on borrowed time, as if I ought not to be alive. Soon after the accident we left the country where it happened, and it wasn't until I was nineteen that I could return. When I did, one of the first things I urgently needed to do was to return to the crossroads where it had happened.

For children who have not had such close escapes themselves, it may be some years before personal awareness of death comes home to them. Even a child of nine or ten will readily be filled with fear that death could happen to him when a neighbourhood child dies or is killed. Such thoughts, though, are usually very quickly put aside.

When a young person comes into the teens, his capacity to deny his own mortality usually melts away considerably; we saw that his new capacity to abstract and also identify more widely allows death to become an intrinsic part of the adolescent's growth of adult identity. The transition is usually gradual. An example of this could be seen in the Second World War. Children in their early teens or less were often blissfully unafraid of the bombing, at least as far as being killed themselves was concerned. The evidence for death was all around them but was consciously denied. By the late teens young people were more often quite openly scared. For instance:

I myself remember becoming frightened of being killed in an abstract way during the war when I was seventeen. But when a bomber dived straight towards me, I remember staring at its black shape, and within a few seconds thought, 'This is going to hit us. Somebody is going to get killed. I might get hurt, but I am not going to die.' My capacity for denial was still working. Three years later exactly the same thing happened again, but this time I was very frightened and thought I was going to be killed. I suspect I was a late developer over this, but something of this transition happens to most people at some time or another.

Nowadays, when immediate death is not present and dying people are usually nursed in private, perhaps this recognition of mortality is more difficult to achieve. However, the question is evidently still present, for young people's songs and talk are often full of ideas about death.

Ideas about mortality come quicker to some than others. Certainly the immediate and personal experience of death is the most convincing teacher. Nurses, doctors, and soldiers in wartime probably mature in this way sooner than others. However, many adults keep up their denial of mortality even after many experiences of danger. The behaviour of some drivers on the roads shows us this by the way they place themselves in absurdly dangerous positions. Perhaps being encased in the familiar shell of their car fosters an illusion of invulnerability.

Even in a country at peace, the evidence for personal mortality is inexorably placed before the individual in his own dangers and the deaths of friends and relatives. By one's twenties and thirties the idea of personal death has usually taken firm root in the conception of oneself. It is an intrinsic component of a realistic sense of one's identity. In it the individual recognizes that he is finite, and is a humble being like all others. It is frightening but refreshing, because a longstanding denial has been thrown off. The feeling of one's life has more limits now and is thus more complex, articulated, deep and interesting.

With this sense of mortality included in his awareness, an individual is often spurred to clarify the things he would like to do in his life, and also to divest himself of dreams that mean little to him or are impossible.

By the age of forty or so, awareness of mortality develops into the sense that life is half over and there is not much time left. Naturally people vary in the quality and time of life of this recognition, but it is sufficiently widespread to be discussed as a general phenomenon in both men and women. It is at heart a *depressive* experience with much anxiety. This is because it is concerned with loss, not necessarily real losses but lost hope or lost dreams for oneself. With a long future in front of him the young man can assuage the sense of his own failures with dreams of the future. But by the forties a person's contribution to adult life has been largely delimited by work and family. The future is restricted both by these and by the physical reality of an ageing body. The old methods of looking to the future to deny anxiety no longer suffice; a new equilibrium has to be found. This is the *mid–life crisis*. Let us consider men and women separately before finally drawing the threads together.

Middle Age in Men

What is said in this section can apply to many women as well as men. The section after this considers questions that apply only to women. Naturally the external pressures on most men in our society will depend largely on their *work* commitment. This has usually been fundamentally delimited when quite young and the nature of a man's mid–life will depend on this early choice. For instance, a young man who has found his identity in an occupation that involves fast physical co–ordination has little to look forward to in middle age unless he can develop to new functioning: he must change large areas of his identity. Such changes are inevitable for dancers, pop–stars, athletes, racing drivers, soldiers, sailors and airmen. The fast co–ordination upon which these depend deteriorates noticeably from the late twenties onwards. Most young people who take up these careers are well aware that the centre of their lives must be transformed in their thirties or forties, and prepare themselves for a two–career life. The second career may grow out of the first, for instance a middle–aged team manager must have been a footballer first. It is not likely to be as glamorous a career as the first, for glamour most often depends on physical articulation which is on the wane.

In the transition from one career to another the individual must *mourn* the loss of his old skill, and at the same time find new investments. If later careers are satisfying, they are, being concerned with the articulation of others rather than the self, often essentially parental and not glamorous in nature, using past experience to contribute to others. This can be seen, for instance, in management, organizing or education.

On the whole, a person who has a highly developed intellect is less likely to be faced by such an acute career crisis as is the physical performer. Intellectual functions deteriorate more slowly than speed of body movement. However, the swift manipulation of concepts falls off in a way which is rather similar to that of physical dexterity. Thus where speed of thought counts, young men and women are at an advantage. For instance, mathematicians, who need to move fast over vast tracts of abstract ideas but need little real experience of the ways of the world, have usually done much of their original creative work by the age of forty. But where speed is less important than the balanced assimilation and organization of large bodies of information about the external real world, the individual probably does not reach his prime until he is forty or more.

Men who have committed themselves to work that requires depth and breadth of experience, particularly in identifying with others, may hardly start being independently productive until they are forty. The same applies to other less intellectual occupations which require steadiness of judgement. A racing driver, for instance, is old at forty, but a bus driver is least accident prone between forty and fifty five.

Something rather similar happens in literature. Poets often blossom young and continue into old age. But many novelists and playwrights only reach mastery of their craft by middle age. One can detect development, for instance, in Shakespeare's style as he grew older. His later plays show much more flexibility and originality in the use of language than his earlier ones. Jaques (1970), in particular, has described these mid–life changes in creativity.

The vesting of authority in older people is not just an outworn quirk of our society. Balanced judgement must rest on the assimilation of relevant information. In many spheres this can be done when young, but in others it may take many years to amass.

This is particularly so where knowledge of people is concerned, for this requires face–to–face interaction with many different individuals. Thus physicians, psychiatrists, social workers, administrators and statesmen probably only reach their prime in middle age.

On the other hand, older people easily slip into stereotyped thought behaviour. Young people are not only revolted by subjection to the rituals of the old but also have life in front of them and are ready to welcome change. Then in exasperation they challenge older people to the familiar generation wars. When this happens the older person, with his settled identity threatened, is only too ready to shrink into a shell of outworn habits.

Although the middle–aged person's slower body must play a part in such a tendency to conservatism, it is probably due more to emotional investment in old modes of thought. In them he seems to cling to the fantasy of the adult identity which he discovered in his youth, and in doing so he often denies doubts about its viability in changing times. That this is psychologically rather than physically based is suggested by the increasing number of people who are able to rebel from their old ways: during the middle of their lives and later. It is not only married women who are fundamentally shifting their way of life as their children grow up. Mature male students also often find that they are keener learners than their younger fellows. Some people become more radical as they grow older, feeling perhaps they, having less to lose, can be freer with their opinions.

When learning anything with freedom and zest a person is humble and romantic. To do this it is necessary to regress and to allow childlike parts of the self, with their loosening of structure and infinities of feeling, to re–emerge. It is a small crisis. An older person often finds this frightening and humiliating, for his self–esteem has developed out of the things he has mastered and he need no longer feel childlike. When an older person can loosen himself and still show this eagerness and use it, he seems to take on an ageless quality. It is as noticeable in some very old people as it is in the young. People having this facility seem to find little difficulty in communicating with any generation. When an older person has kept the child alive in himself, he can not only empathize with younger people but also learns from them, and in this way is also naturally acceptable.

Middle Age in Women
(Giele, 1982)

Some of the main problems of middle age for women have already been discussed in Chapter 12 on parenthood. There we stressed the plight of mothers who, having been engrossed in their families, find themselves, with much life ahead of them, largely unneeded by their children and financially subservient to their husbands. Whereas many women find this pleasant with plenty to do, for others it is deeply unsatisfactory and they strive for a new work identity and hence a changed relationship with their husbands. It must be an appalling failure, either personally or by the forces of a society, if a woman cannot be free to do this.

Compounding this is the menopause (Phillips and Rakusen, 1978). Its hormonal changes evoke disturbing symptoms, hot flushes, dizziness, difficult and irregular periods. Even more than the symptoms themselves, they mark the end of childbearing and impress on a woman that her life is irreversible and finite.

Uncertainty must pervade such a time for everyone, for no one can be sure whether the change will bring a dried–up sexual decline or settling enjoyably into a new springiness. It may be that many of the disturbances ascribed to the menopause are also symptoms of the older woman's identity crisis. For instance, many people have pointed out that having successfully brought up children and then having found the enjoyment of a new identity, the menopause and after is not particularly depressing. Its end is even a new freedom as there is no need to worry about getting pregnant any more. Although it has been known for centuries that sex need not end with the menopause, it is still a new discovery for many people.

Naturally this affects men too, for husbands often readily slip away from sex if their wives have gone off it. When this happens a man's virility seems no longer to be centred on his penis; he often seems to turn paunchy, taking on rather old–womanish characteristics. Women also seem to lose the centredness of sexuality in their genitals. When this happens they often develop rather heavy, even mannish 'battleship' features. The sight of other men and women of the same age maintaining their youthful vigour suggests that this disintegration of sexual centredness is not a biological necessity but rather a psychological phenomenon.

Women who are particularly disturbed by 'the change' seem often to be those who have not found their way to conceiving of anything else to do except rear children. They are dissatisfied and frightened of the future. This is then a serious identity crisis, not just the menopause.

However, another group of women who do more directly feel deeply depressed about the menopause are those who have never had children. To them it is a time of real loss of biological fulfilment and sadness. It is not the same either socially or psychodynamically, as the identity crisis just mentioned.

Let me end this particular discussion with a word of warning. We have stressed the importance of the identity crisis rather than the menopause itself. Even though this crisis is not necessarily caused completely by the change of life it is still fundamentally linked to biological ageing. It is part of the endless cycle of life and death. What is more, understressing the physiological menopause may encourage the attitude that it does not matter. Quite a lot of people brush it aside denying the underlying feelings they have. It must affect every woman and always have a depressive note of loss to it which is best to recognize. If recognized the menopause is sad but can also herald new freedoms.

Middle–Aged Parents and their Children

The dramatic problems experienced by parents with their adolescent children have often been noted by psychiatrists and others. We saw in the discussion of adolescence that parents themselves often seemed to need to go through a reworking of their own adolescent questions, in order to break their old preconceptions and find their own beliefs rather than parroting outworn ones. Parents must actively and consistently show their children into the adult world. To do this they seem to need to use their own adolescence as a starting point.

Not only are middle–aged parents' beliefs explicitly challenged by younger people, but they also have many distresses of their own which are aroused by their children's youth. Parents' envy of their children's vitality is one. This often seems to arise when a parent feels he was cheated out of a full adolescence himself, and then resents it in his children. This can be compounded by the menopause where a mother becomes preoccupied by her waning life and by not having a second chance.

Less often mentioned but equally insistent is the converse of this envy. Here parents seem to be tied to their children, preoccupied because they feel, realistically or not, that they failed to give something which was vital to their children's upbringing, and their children will be incomplete without it. This is often most painfully felt by parents of handicapped or disturbed young people. They feel they have not completed their job properly.

Sometimes this is clearly a neurotic fantasy, but for parents of handicapped children it is a very real misery. They often can never have the satisfaction of letting a child go with the assurance that he will manage on his own. Detailed discussion of this is outside our province, but the question does highlight that it is not only children who want to grow up and away from their parents. Parents, when they are enjoying themselves, can really be glad for their children to grow away from them. Not only has a burden been lifted, but they have the pleasure of seeing that they have done a job well enough. Their child's success is theirs too.

While we are talking of success, our discussion of the *depression* of mid–life has not yet touched upon its opposite; the effect of elation and the manic moods of parents. Mid–life and onward can be a time of profound pleasure in fulfilment, of success for a parent, particularly some fathers. They may be honoured by people around and revel in knowing many influential people. Such grandiosity can be very difficult indeed for a young person, especially if his parent is intolerant and opinionated as well. The young person is then likely to feel he can succeed in nothing in his parents eyes, he may even turn to punishing them malignantly by chronically failing himself.

The Mid–Life Crisis

This is a term in general currency given to those crises which, in at least an underlying way, arise from a person's awareness of the ageing of his body and of the passing of half his life. It has been particularly stressed by Erikson (1963) and Caplan (1964); a very searching description is given by Jaques (1970).

It is stressed by these writers that mid–life is at root a normal crisis arising out of awareness of ageing of the body and growth of a younger generation. This awareness may be avoided by all sorts of defensive tricks but change in old ways of life must occur for better or for worse sooner or later. As in any crisis, chronic pathology does not arise inevitably out of anxiety or depression. Rather, it arises either when a person frantically clings to old patterns of thought and behaviour which are no longer viable, or, having given these up, he can find nothing new to take their place and sinks into regressed despair.

The crisis is essentially concerned with the experience of *loss*: loss of hopes, loss of old, enjoyable abilities and loss of capacity to procreate. Mid–life is likely to have profound satisfactions too, but loss is omnipresent. Its progress is thus marked by *mourning* and its resolution is found in *renunciation* and *sadness*. Feelings about death are never far away. This is partly because it is stirred in the first place by recognition of one's own future death, but also because it concerns the death of old hopes and pleasures. These in themselves re–arouse unconscious, primitive childhood experiences of loss and distress. With these come fears of the shattering of the self (as described in Chapter 4), which is a dreaded 'death of the self'. Fundamentally it must assail both men and women but, as we have seen, owing to differing physiology and social pressures, the timing and forms it takes tend to be different for both sexes.

Jaques has pointed out that it is essentially a *normal depressive* crisis, whereas adolescence is a *normal schizoid* one. He makes this differentiation to highlight that mid–life is a crisis of *loss*. In adolescence, losses occur but they are not the essence, rather it is a crisis of *finding* an adult integrity of the self. Because differentiation of the self is at its essence, with the psychotic anxieties that this entails, it is called schizoid.

However, although this fundamental distinction is, I think, unassailable, we should also remember that the mid–life crisis can involve finding, partially at least, a new identity. This is particularly so for many women at the present time. In which case, many features of what is often thought of as 'adolescing' occur in mid–life.

It is for this reason that we stressed adolescence as a model for future crises and argued that 'adolescing' does not necessarily

248

finish for good in the early twenties. Many aspects must be reworked as new identities are found.

Jaques (1970) has also pointed out that mid–life is not just a time of loss and renunciation but rather of change. He suggests that an examination of works of art and literature indicate a movement of style from the lyrical one of early adulthood, through tragedy, to a more serene and rounded one which characterizes later life. At the same time, mid–life, being a time when many people are particularly aware of tragedy, is also one of urgency. Desires which have hitherto been vague must be brought to fruition. Thus middle and later life may be a time of great creative energy and enjoyable sense of purpose.

It is a common assumption that the impatient innovators of change are all young. This may be true when considering people in the mass. But the organizing centres of innovation and change, either socially or in ideas, are very frequently individuals in their later life. Taking extreme examples, the major political revolutionaries of history have mostly been middle–aged at the time of their fruition. Lenin is a case in point. Britain's major revolutionary organizer, Oliver Cromwell, is another outstanding example. He was probably fired with Nonconformist zeal as a boy by his schoolmaster. But it was not until he was in his late twenties that he became deeply committed to religious puritanism. In his thirties he became politically active, and he was over forty before he moulded himself, for good or ill, into a military leader and vital political organizer.

A more peaceful but poignant example of the recognition of mortality as a spur to self–organization can be seen in Guiseppe di Lampedusa. He was an Italian nobleman who, though well known as an intellectual, wrote nothing of great note until his late sixties when he was told that he was incurably ill. Spurred on by this news he set about writing a novel which had been planned for many years. *The Leopard* was published after his death, and was immediately recognized as a modern classic.

FURTHER READING

1   Anthony, S. (1971), *The Discovery of Death in Childhood and After* (Harmondsworth: Penguin). Just as its title describes; interesting even enjoyable reading.

2 Bromely, D. B. (1974), *The Psychology of Human Ageing* (Harmondsworth: Penguin). A summary of research findings about the ageing process from mid–life onwards.
3 Caplan, G. (1964), *Principles of Preventive Psychiatry* (London: Tavistock). An original work which expounded the concept of crisis, in mid–life and in other phases.
4 Jaques, E. (1970), *Work Creativity and Social Justice* (New York: IUP). Of particular note for his essay 'Death and the mid–life crisis'. Its language is psychoanalytic, but easy reading and a classic of its kind.
5 Neugarten, B. C. (1968), *Middle Age and Ageing* (Chicago: University of Chicago Press). A technical, psychodynamically oriented study on the subject, but of general interest.
6 Phillips, A. and Rakusen, J. (eds) (1978), *Our Bodies Ourselves*, Boston Women's Health Book Collective (Harmondsworth: Penguin). A health book by and for women of all ages, but good sections about the menopause.

CHAPTER 15

Old Age

Respect for the Old

In a static society we might expect the old to be respected. Having lived longest, they can act as the most assured carriers of techniques and beliefs which may be passed on virtually unchanged from one generation to the next. This reverence for the old certainly occurs in many societies, especially if they have both tight family organizations, needing an old person's authority, and a plentiful food supply for those who are too old to be producers themselves. However, where food is scarce and living conditions are hard, as for instance, with the Eskimo in times past, the old become a burden upon the young and may expect to be discarded.

The conditions of our society seem to bear some resemblance to both these forms. With a prolonged expectation of life and a highly developed economy, a large proportion of the population is aged and apparently unproductive, yet can be supported. At the same time, old people are not seen as essential carriers of culture. Their contributions are indeterminate, so that very many of them have the appearance of being useless and unwanted.

The ageing person is presented with threats to his well–being from two directions, firstly from the deterioration of his bodily functions and secondly from social expectations that he will become useless on account of his age, irrespective of his real capacities. Many people are compulsorily retired at a fixed age, not for their own benefit but so that the younger generation can step into their shoes, this is particularly so in times of high unemployment. There are also vague but widely influential attitudes which match this compulsory retirement; for instance, an old person's ideas tend to be dismissed as old–fashioned simply because they come from a person of aged body. He may be physically cared for, but is not reverenced and frequently not even respected.

The Crisis of Retirement
(Bromley, 1974; Burstock & Shana, 1976; Neugarten, 1968)

Just as middle age was perhaps epitomized by withdrawal from intimate parenthood, so, for many people, old age is marked by retirement from money–earning work. This is another crisis of life. New integrations of functions must develop for life to continue fruitfully and in contentment. Retirement from work is only one facet of ageing, but it is clear–cut and easy to recognize. Similar, smaller retirements and withdrawals are occurring throughout old age, and likewise call for internal development and reorganization. We shall stress retirement from paid work here simply because it epitomizes much about this time of life.

A wage–earning man or woman's identity or sense of self will have been largely oriented around working capacity. He may or may not have invested much of himself and his imagination in it, but will inevitably know himself by the work he does. With retirement this identity is taken away from him. So also is a good part of his income. Here are a few examples of how this crisis is managed.

Mr H said he had mixed feelings when retirement was approached. He looked forward to not having to work, but wondered what he was going to do with himself. His wife said she did not want him hanging around the house all the time. On the actual day, he went to work as usual and wished his friends good–bye at the end of it. They settled down to being pensioners. The greatest readjustment was living on a reduced income, but the first ten years proved to be happy. They were both healthy. Mr H took an allotment, so they never had to buy vegetables. They lived close to their children and helped to look after the grandchildren. In the summer Mr H returned to work part–time, but in the winter they could relax with their friends and social activities.

Mr I says he enjoyed every minute of his life as a mechanic. But he looked forward to retirement and, although offered a number of part–time jobs, turned them all down because he felt he needed a complete break. During the first months Mr and Mrs I had a series of holidays visiting children and grand-children and two weeks away on their own.

After this Mr I turned into a Jack of all trades. He moved around on a bicycle doing gardening, decorating and small repair jobs. His evenings were taken up with committee work. His health was good, but he noticed that there was not so much spring in his legs, and now needed a rest after two or three hours' work. His memory was also not so good, and it took a little longer to think things out. He might not do as much as he used to, but the quality of his work remained satisfyingly high. He also had a freedom which was denied him when in employment.

In contrast here are two other people who did not develop new patterns of living and despair set in.

Mrs F went on working in a pub until she was over seventy, when she slipped and bruised herself one day. She lost her job through being off sick. She recovered quite quickly, but never went back to work. Her husband remained lively and full of jokes, but she slowly became quiet and apathetic. She complained of aches and pains, but managed the housework and cooking. Then she began to lose weight and looked drawn, gave up doing her hair and using make–up. She usually shuffled about the house in boots and several layers of clothing.

Mr G had been a pig farmer for most of his life. He never gave a thought to retiring. By the time he was over seventy there was a noticeable deterioration in the farm. Roofs were leaking and fences broken. Mrs G realized her husband could no longer keep up, and suggested he employ a farm–hand. But he was too stubborn to admit his declining abilities and would not listen to his wife. Finally he became acutely ill and was taken to hospital, where he was overcome with despair as he lay in bed. For many years he had dreamt of his family buying up the land around and continuing in his tradition of farming. But he had no sons, and his daughters had all married townsmen. When he dies he will be the last of a long line of farmers.

What functions come into play in this crisis? We have already noted the losses from physical deterioration and attendant depressive feelings. As with any loss, this usually sets in train a process of *mourning*. Here fear, rage, paranoid ideas and regret may be followed by finding substitute satisfactions and then peaceful sadness at the change.

On the positive side, the crisis involves finding new investments. One particular psychological quality seems to be important here. We saw how children approach new activities with awe and enthusiastic romanticism. If a person has kept this childlike quality as part of his functioning identity, he is likely to be able to use it even late in life to discover and develop new investments.

However, such internal changes can only take place when the external social situation presents opportunities for new investment. For instance, Mr I the mechanic lived in a fairly built–up area, so that his leaning towards becoming a general handyman met a local need. Mr G the pig farmer, on the other hand, could not have turned to this even if he had wanted to, because there were few houses in his rural neighbourhood.

The enjoyment of old age seems to rest, as it does throughout life, upon fusing the satisfaction of personal, selfish urges with contributions to others. The environment delimits both of these as much as does an individual's psychological character.

A great pleasure of old age is that few responsibilities are now expected of a person, and self–centred pursuits can be enjoyed without recrimination. This absolution from guilt must lie at the heart of many old people's enjoyment of their lives, which seem sparse in other ways. Remember that many older people really do enjoy their lives, however odd this may seem to young people. Perhaps the most common handicaps to carefreeness are lack of money and health, or fear of this lack in the future. Anyone acquainted with old people will have encountered real poverty as well as ill health.

Just as vital to many individuals, especially in early old age, is the desire to continue *contributing to others*. Not only does this give real happiness to everyone involved it also keeps the rigidity of old age, at bay. *Keeping in touch stops you going senile.* Our present social situation is such that although people are tending to lead longer lives, they still have single careers, so that the last quarter–century of a life may be economically unfertile. This is perhaps slowly being recognized, and agencies for employment in part–time work are being set up. Perhaps in the future a life of several different careers, each attuned to age, will become a normal part of our social order. We must perhaps be sceptical that this will come easily as so many vested interests conspire to have the old out of the way.

The break up of extended families living in close proximity has lessened the opportunity for old people to help in child rearing. But this does not mean that the need for such help has disappeared. It is a great strain for a mother to bring up children without assistance either from grandparents or other help. A mother who has no grandparents or aunts to fall back on when needed is often chronically worried about exhaustion or illness. The illustrations given earlier in this chapter show how both old men and women may turn to being auxiliary parents. It is often noticeable that particularly affectionate bonds grow up between children and their grandparents. Grandparents are not burdened with the parents' anxiety and responsibility, so that they can be both more patient and have more time to relax and play in a childlike way with children. Communication and a sense of equality often grows between the old and young which makes them very fond of each other (Brubaker, 1983; Burstock & Shana, 1976; Townsend, 1963).

Many other activities readily spring to mind to which old people may contribute. We have already noted paid work, odd jobs and committee organization. The more subtle functions arising from ordinary informal conversations seem none the less important. For instance, old people are the living *historians* of a society. By personal reminiscence they tell younger people how things were in the past, both in their own lives and those of their parents and grandparents. Thus their span of personal memory may bridge a century or more. A young person may not be called upon to copy their ways as in a traditional society, but historical conversations help to impress the reality of his place in time. He is made aware that he has grown out of people before him, and that others will come and take over from him in turn.

More generally, an old person, having experienced many things and being close to death, can often express himself about living with a cool, dispassionate clarity devoid of the grandiose illusions and denials to which younger people are prone. With this natural wisdom they can broaden our minds and make us humble. They can tell young people, 'This will happen to you one day'.

Depression and Ageing

The common feature of ageing is, of course, shrinkage of body functions. Cell tissue is not replaced as in youth, so that hair thins, skin and muscles lose their vibrance, blood circulation malfunctions and brain cells die. In the last chapter we discussed how, from childhood onwards, there is an impact upon awareness when body parts become diseased. With ageing 'small deaths' become widespread, and moods of acute anxiety and depression arise; this happens at any time of life whenever there is a loss of enjoyable functioning. With old age there is less opportunity than before to hope for the future, so that a valuable defence against depression is no longer effective. Coupled with physical deterioration comes the loss of friends and loved ones who will also be ageing and dying, and this also brings depression.

Many individuals accept the shrinkage of their lives with resigned equanimity. Perhaps one factor in this is that loss of vitality can act as a release from guilt. The inner commands of 'ought, must and should' may lose their grip now that it is no longer physically possible to carry them out. The old person is in the position to relax into unambitious contentment. Perhaps those people who have been forced on many previous occasions to accept their fate in a passive way find this contentment of old age most easily. On the other hand, those who have characteristically dealt with uncertainty by self–assertive activity may feel the losses of age more acutely. But no matter how it is received, ageing seems to be inevitably attended by deep and widespread depressive feelings. These can vary from utter misery to quiet peacefulness. It may sound strange to refer to depressive feelings as peaceful. However it is well recognized that, after the first depressed pain of the experience of a loss, acceptance can lead to a dignified sense of peace which can be very happy.

Persecutory and Paranoid Feelings in Old Age

Physical deterioration means that there is a concomitant depletion in ego–functioning and shrinkage of the sense of self. The individual feels smaller and more helpless, so that the outside world is consequently larger and less controllable. Events in the external world tend to be experienced passively, and hence may become more persecutory and threatening. This is manifest in an old person's natural fears of roads, trains, and whirl and bustle generally.

As ego–functioning and the sense of self tend to shrink, so we would expect primitive, less integrated patterns of thought and feeling to emerge. This regression is manifest not only in an old person's generalized fearfulness, but also in his tendencies to use projective mechanisms. In the prime of life, these are often held in check or elaborately covered by rationalizations. With ageing, these defences often fall away so that projections emerge more nakedly. This is most evident in senility.

But even a person in full command of his faculties can slowly subside into paranoid cantankerousness. For instance:

A widow had had a very active life. She reared a large family, and was also respected throughout her town for being generally interested in people and helpful. As her body slowed up, so she began to become bitter and suspicious. The first sign was an argument with neighbours over a dilapidated fence. This grew into bitter resentment, which spread to anger at all the people she had helped in the past and who now seemed callous towards her. She would, however, let no one near her to help, because she was convinced that they would make matters worse. Perhaps when young this woman had overcome dread of the passive feelings of being helpless and uncared for by developing a very active and rather bossily helping character. When this activity was no longer possible, her sense of persecution and helplessness emerged and consumed her whole attitude to life.

Another instance of projection is commonly seen with regard to awareness of physical deterioration. Thus the sense of body malfunctioning, 'little deaths' as I have called it, is often projected on to the external world, so that dreadful things are felt to be happening outside rather than inside the individual. For instance:

An old countryman was looking at the thatched roof of a cottage and said, 'That thatch is good for another ten years. I'm eighty and never had a day of illness in my life, but the thatch will outlive me, no doubt.' He then lapsed into projection and said, 'But then such terrible things are happening in the world now that it will all be over for everyone on this earth before then, that's for sure.'

Such collapses into paranoid feeling are common but not universal. They are most marked when physical functions have deteriorated and little activity is possible. There is much more to old age than this, but we have discussed deterioration, depression and persecutory anxiety because these are underlying background feelings in ageing. They have to be dealt with by every individual as he grows older, however fit and active he may be.

Dependency upon Others

We have so far been considering the position of old people who still retain their vigour. So long as a person can think and move, he maintains his independence of judgement and hence his individuality. But the progress of shrinkage is inevitable, and the old person has to sink into dependence upon others. This must be painful for all concerned and is often bitterly resisted.

Dependence provides the ground for neurotic satisfactions for both old and young. This is often evident between old parents and their children. For instance, an old person losing his independence of action can gain malicious pleasure in being a ruthless baby who will never be satisfied. The impulse to behave in infantile ways may have been grudgingly held in check for many decades from the days of the old person's own childhood, and only now given vent. The younger person, on the other hand, may take pleasure in holding a parent helplessly at his mercy. Perhaps having felt subservient all his life, he can only now wreak both revenge and also appear virtuous in taking up a succouring stance. There are many other ruthless games played between old and young. No doubt fantasies like these play through the minds of everyone. However, they only become ruthless or neurotic when fantasy invades and blots out each person's recognition of the other's individuality. It may also be noted that such games are not confined to parent–child relationships. They are readily detectable in some professionals working with the aged.

Where mutual recognition of each other's dignity is maintained, underlying fantasies will inevitably still be active. But they will be held in check and used as a spur to activity rather than being a dominating force. When this occurs, care of the aged may still be painful and sad but also deeply rewarding, for both old and young are together close to the elements of life and death.

Senility

The progression into muscular incapacity is deeply depressing, but deterioration of brain function is perhaps worse because communication then becomes transient and difficult. With brain deterioration the usual modes of conversation disappear, so that helpers are left embarrassed and confused. With the lapse of highly integrated thought process, more primitive mental functions emerge unfettered. In particular, the paranoid projective mechanisms mentioned earlier often dominate a senile person's thoughts and speech. These are often directed at helpers who can be made to feel hurt and confused.

Bitter paranoia does not occur with every senile person, but all seem to revert to primitive modes of symbolic thought and communication. Conversations take on 'unconscious' qualities often only known otherwise in dreams or psychosis. This regression is disturbing to helpers who are naturally attuned to ordinary objective speech. The primitive symbolism of a senile person can often, however, be deciphered by a listener. When this is achieved, the contact between him and the old person can be very moving for both.

The code of this primitive symbolism, being like that of dreaming, can be roughly described as one where *generalized abstract ideas are expressed in terms of specific body functions, and conversely body functions are expressed in terms of generalized or external events.*

This is the 'language of the unconscious'. We have discussed it, all too briefly, in the early chapters of this book. There we saw a strange phenomenon in primitive imagination; particularly in dreams, neurotic symptoms, psychosis and in the background of little children's thoughts. This is that an intellectual concept about something can be experienced as *identical* to a bodily function if they have *a similarity* of form in common. (Matte Blanco, 1975). Now we have it naked in senility.

Listening to such equating of mind and body can be very disturbing, but when a listener has mastered the code and knows the old person he can then often carry out the necessary translation, so that he understands what is being said.

Here is an example of an abstract idea being expressed in terms of body functions.

An old man was clouded and wandering in his mind. His son came to ask for his signature for a power of attorney to manage affairs, and was worried whether his father would understand and comply. The old man seemed to wander off to another subject when he said, 'I must now climb up on to your shoulders. I hope I won't be too heavy.' The son, realizing what he meant, said, 'Oh no, everything is in order and we shall be able to manage things quite well,' and then handed him the pen and paper. The father signed it and with a sigh said, 'That's much more comfortable now, but then you always were a broad–shouldered boy.'

Here is an example of a body function being expressed in terms of an external general event.

An old woman was found crawling in the corridor whimpering, 'There has been an earthquake.' The nurse looked and found that she had soiled herself.

We can see here how external and internal reality have become confused, as may happen to all of us when waking up from a dream, or like a person in psychosis. A young baby has also probably not yet made the differentiation between self and external reality. Such differentiation is one great task of early childhood. In adult life we use our inner fantasy to feed our creative imagination in relation to the outside world. Then with senility the differentiation breaks down, and we can communicate only through concrete symbolic language. Perhaps as people become more sensitive to primitive symbolic communications, the loneliness of old and senile people will be somewhat alleviated.

FURTHER READING

1 Bromley, D. B. (1974), *The Psychology of Human Ageing* (Harmondsworth: Penguin). A wide summary of research findings about the aged. Invaluable for finding more detailed references.

CHAPTER 16

Dying, Grief and Mourning

Industrial society has, maybe, enhanced the possibilities of individuality, but a price has been paid in loneliness, particularly for the mother of young children and in old age. Dying, too, tends to be hidden and often solitary. Ours is a death denying Society (Kubler–Ross, 1975). Gorer (1965) among many others has amply demonstrated this.

In close–knit communities we would expect a dying person to be surrounded by people who know him; death may then be a public, shared event in living. In some rural areas of the British Isles particularly Wales and Ireland this is still not uncommon. Many local people help in nursing. After death it is often assumed that any acquaintance old or young can pay their last respects to the body laid out in a bedroom or the parlour. In areas of mobility such public intimacy is most unlikely; it is unthinkable for the inhabitants of a suburban road to crowd in to see a body, for most would be strangers. What is more, modern medicine means that many people die in hospital, often alone except for the staff. Sudnow (1967) and Bok (1978) have shown that within a hospital the staff themselves develop social ways of coping with death, which often continues the isolation of the dying person.

Death is frightening from childhood onwards and taboos grow to keep the anxiety at bay. In Britain, these taboos, in the recent past at least, seem to have been directed at hiding death away to allow the living to deny its omnipresence. Younger people then have little direct experience of real death and the personal humility it can bring. The old and dying for their part tend to be left isolated and alone. Thanks largely to the work of such writers as Hinton (1967), Gorer (1965), Kubler-Ross (1972), Saunders (1959) and Schneideman (1984) perhaps this taboo is beginning to melt.

Apart from the immediate considerations for the dying, we all need to be brought to awareness of death. For it is only by this can we fully appreciate living (Kavanagh 1972).

Terminal Illness

Here is a description of the last weeks of an old man. He was very much of an individual, so his way of growing old and dying was his own, but many of the features are common to others' experiences. (He was not known to me, and I am indebted to Barry Palmer for the description.)

For four years Mr J has been partially paralysed and was unable to move outside his home except in a wheelchair. Even so, although eighty seven years old, he was very alert and able to converse on many subjects through his daily reading. His favourite topic of conversation was, nevertheless, his past career as an officer. He was very proud of being an active man.

He had steadily deteriorated in physical health, and on many occasions it was thought that he might die. But he was never willing to admit that his health was declining and, when asked how he was, would say that he was better, and even demonstrate this by getting out of his chair and walking across the room. Each step might take upwards of ten seconds, but it would satisfy him that he had convinced the onlookers that he was not dead yet. His will to live was extraordinarily great. Mr J was afraid of being forcibly admitted to hospital. On many occasions it was advisable for the sake of his wife's health that he should leave home for a couple of weeks. He refused on every occasion, and one felt that this was because he had the fear that entry to hospital would be the end of him. On one occasion his wife had to go into hospital, and in spite of the fact that he was most uncomfortable and not properly cared for, he would not leave the house. Mr J was always conscious of death. He was very fond of his children, but if they called without warning he would get very heated and suggest that they had been sent for. I do not know whether he was afraid of death, but he was of losing command of his life situation. It was always clear that he wished to be the boss of the house and in command of every aspect of his life. The thought of losing his independence, which he had largely done but would not admit, was very apparent.

Three weeks before he died Mr J was walking to the toilet

when he began to slip, and eventually sat helplessly on the floor. Normally he would claim that he was all right, though someone would have to help him to his feet. On this occasion he said, 'Joan, I'm going.' He did not say much more, except that he wanted to see his children. From this point on he was no longer the man who wished to appear in command, but allowed himself to be cared for. A few days later, having seen his children, he became unconscious much of the time. The one thing which differed from his former collapse was that, although it appeared much the same to the onlookers, he knew that this time he was going to die.

This was one man's way of dying. Although there have been countless millions of deaths in human history most of us in the urban West really know very little of what it is like. This is not surprising because, apart from our fears that make us shy away, it can be very private, silent and alone, for dying people are often not much given to talk. It is perhaps only nurses, even more than doctors, who have a wide experience of the last weeks of life.

There is doubt about when an organism can be said to be dead. Some argue that absence of cerebral activity is decisive, others that absence of pulse and heartbeat is, but for our practical purposes this is an academic question. What happens before death and after does concern us. In traumatic instances death, of course, may be instantaneous. But in other cases a terminal phase can usually be recognized. It may last only a few hours in, say, a coronary thrombosis or be drawn out over weeks and months in slow–working diseases like cancer (see Veatch, K. M. in Schneideman, 1984).

The terminal phase has begun when a disease or deterioration of tissue has such a hold that it is irreversible. Those around may not be aware of this, but there is some evidence that the grip of the lethal process is signalled to the mind of the dying person. For instance, it has been noted that the quality of dream imagery changes. We know little about this as yet, however Hinton is convincing when he suggests that dying people are usually fundamentally aware of their dying, even though it may be dim and often denied.

The months after the diagnosis of a fatal illness are usually quite awful, lightened often, for us around, only by the sheer courage of the dying person. Thanks to such workers as Kubler–Ross (1975) the thoughts and ideas of people who have only a short time to live are much more known to us than only a few years ago. She has stressed how much work there is to do in a short time to prepare oneself to complete one's life. She has called it *the final stage of growth*.

It has been pointed out that a clear emotional sequence can show itself in the period following diagnosis of terminal illness (see M. Imara in Kubler–Ross, 1975). First there is shock and *denial* of the illness followed by loneliness, internal conflict, guilt about and sense of the meaninglessness of one's life. This turns to *anger* at one's fate followed by bargaining with fate or God. Then comes a gradual realization of the real consequences of dying leading to *depression*. After this trough there is a movement of increased self–awareness and enjoyment of contact with others giving on to *acceptance* of dying with self–reliance.

Such a sequence can rarely be clear cut and not many people work through, or 'pain' through to equanimity. However, it is not as rare as all that.

One of the worst experiences in this period seems to be uncertainty, 'how long have I to live', 'how long will my mind be clear'. With such indefiniteness preparation is very difficult. It is not uncommon to hear, 'I'm ready now, the pain and helplessness isn't worth it any more, I wish they'd let me die, what are they keeping me alive for, not for me I'm sure'.

Physical pain is frequently a person's greatest dread. Many dying people do experience chronic pain and as a direct result comes exhaustion and often deep depression. Pain is often made more acute by anxiety. Perhaps because of this the old tend to feel less pain than younger people. A younger person is less prepared for death, has more to lose and often is distraught about dependents left behind and life–work uncompleted. Nurses often say that the most harrowing experiences are caring for young children, and parents of young children who are dying.

Almost as dreadful as pain is chronic discomfort and immobility. This, too, is terribly depressing. So also is guilt and shame at being a nuisance and incontinent in front of others.

These are probably all at their worst in the earlier stages of terminal illness when a person is still very aware of other people and attached to them. During this time there often seem to be

265

many *memories to be relived*, and settled in the mind. So also are there bonds of affection to things and people to bid goodbye to. This is rarely done explicitly in an orderly manner, but it can be often seen and heard happening none the less. Then, perhaps a day or so before the end, maybe longer, a person seems often to release his interest in those around him. He withdraws into himself; he may talk a bit to those around, but they do not matter to him much any more and he seems to be waiting to let death take him. He may become unconscious, but even if this does not happen a subtle change takes place so that those around usually recognize that death is very close.

Those around may well recognize that the dying person still has work to do. In some aspects of this, especially in bidding farewell, they can help him actively. In other aspects he needs to be quiet, alone perhaps, or with another person in the room. It is helping a natural process, which dying is.

Many people around will often be more frightened and frantic than the dying person himself. With their flurry and tricks of mind they can often make dying confusing and more painful and lonely than it need be (Bok, 1978; Sudow, 1967). People quite often say to a considerate attendant, 'I know I'm dying, but don't tell my relatives they'd be too upset'. This must be the fruit of the taboo of death, so common in our own culture, yet hardly known in many others.

Awareness of death and its recognition by those around can mean that a person ends with dignity and peace, and will be remembered for it with gratitude by those remaining. Here are three instances:

An old man in considerable pain said to his son, 'When this nonsense is over you will find the will in the desk downstairs'. He then proceeded to give instructions about his affairs.

An old man said to his niece, 'I think I shall be gone by the time you visit me again next week–end, so I think we had better say good–bye now'.

An old woman in Nigeria was nearly ninety when she fell ill. However, she said to her children, 'Don't waste your money preparing for a funeral now, because I am not going to die yet'. She did not die, but one day some months later she got up and began to make her preparations by walking round to all her friends saying good–bye. The next morning her daughter was due to go to market and the old lady said, 'I think we had better say good–bye now before you go'. A few hours later she died.

After this the living person changes into inert, shrunken remains. What is left alive of him on this earth now lies in the minds of other people. His relatives, children, friends and acquaintances, all those who have experienced him in some measure, probably have learnt from him and have something of him inside themselves. We shall now turn to these survivors, for they have the task of transforming their experience of the dead person, from someone outside themselves into an internal, living and usable memory.

These are all old people. The dying of a child is far more dreadful, for those around at least, to bear (Bluebond–Langer, 1978). Just as painful, even more perhaps, is the death of a parent, a mother particularly, of young children (Berlinsky & Biller, 1982; Furman, 1974; Pepper & Knapp, 1980).

Grief and Mourning
(Langer, 1957; Lindemann, 1979; McNeill Taylor, 1983;
Parker–Weiss, 1983)

Here is a description by a woman of her experiences of death.

When I was thirteen my grandmother died. She was a much–loved, very elegant lady, with long, black skirts and button boots. I thought as a very young child that she must be a close friend of the Queen. One Saturday she decorated my bedroom, putting on wallpaper that I had been allowed to choose myself for the first time. She sat eating her tea with my new puppy trying to bite the buttons off her boots. She was laughing and happy. I slept in my newly–decorated bedroom which smelt of the new wallpaper. That night I was very conscious of how much my grandmother loved me. On Monday morning an aunt came round to our house, just as I was leaving for school. It was the day I was due to take an English examination. My mother and aunt tried to make me go to school, but something in their faces terrified me and I refused to go. Eventually they told me that my grandmother was dead. She had died very suddenly the previous night. I then went to school, sat the examination and did not think about my grandmother. I did not think about her again until the funeral, when I sat in the carriage with the hearse and the coffin covered in flowers, surrounded by relatives, dressed in black, and I had a black coat too. I did not think about my grandmother herself, because I was too busy trying not to laugh and laugh, and I knew that if I did nobody would understand why, and my mother would be ashamed. When the results of the examinations came out I was top of my class. It seemed as though this was the most shameful thing I had ever done.

A year later my school–friend died. She had pneumonia. I had seen her and she was getting better. The next day my mother told me when I came home from school that she was dead. I did not believe her. I went to see her mother, who told me that my friend died because the doctor would not wake her up to give her medicine. I was very afraid. I dare not go to sleep that night in case I should see my friend. I had loved her, but her death had turned into a terrifying horror.

When I was fifteen my grandfather died. He was ill for a long time, and we knew for a week that he was going to die any day. Perhaps he wanted to die because my grandmother was dead. We had an end–of–term school dance on the Friday, and all I could think of was that I hoped he wouldn't die until that was over, or I should not be able to go. He died on Friday night whilst I was enjoying myself at the dance. I hated him for dying then, and became even more afraid of death.

My next experience was years later, when my own daughter was fifteen. Normally she returned home from school with three friends, two boys and a girl. This particular day she was delayed at school and they left without her. A thunderstorm broke as they came across the common. They sheltered under a tree with their bicycles, lightning struck them and all were killed. I could not find words to console my daughter. I could only be thankful that she was alive.

A little later my father died, driving his car home from a football match. My mother did not ring me up to tell me until two hours after his death, because she 'wanted me to have my tea first'. I was kept very busy looking after her during the next two weeks, and hardly thought about him.

When I was forty four my mother died. She insisted on living alone. She was very independent and had enough money to do so comfortably. I was always afraid she would have an accident and lie unable to get help. She had a minor stroke and lay unconscious for a whole day before I could get there. I did not say good–bye to her. My aunt said she didn't tell me that she was ill. She thought I might be upset and I 'couldn't do anything anyway'.

A few months later my dearly loved father–in–law died. He died of cancer and we knew that the end was near. We received a telephone call to go to the hospital. Our car was caught in a traffic jam and we arrived at the hospital five minutes after he died. We never talked of any of these people and their deaths.

The last one was my husband and he died after three months of torture. He had cancer and refused an operation, which would have been useless anyway. I used to pray that he would hurry up and die, and was glad when he did.

After they had taken him away his dog ripped up his bed and his pillows and scattered the feathers all over the stairs. I changed all the furniture around the day he died, and then I carried on doing all the same things that I had done before. I used to pretend he was at work.

I couldn't stop buying the foods he liked from the grocers. I didn't know when to go to bed or when to get up. My son drove his father's car, and the dog nearly went mad every day when the car came down the drive. We sold the car. Eventually we had to find a new home for the dog, who would have died of misery if we hadn't. My friends were all very kind. They kept me very busy with empty business. I was lonelier with people than without them. Nobody could talk about my husband. Christmas was hell. We always wound up our special clock on Christmas Day. I let it stop. My birthday, our wedding anniversary, his birthday and his children's birthdays were hell, but we never spoke about him. After a few months people started to say how brave I'd been, and what a wonderful new life I'd made. There was relief in their voices. They could look me in the eye again and didn't need to be embarrassed. I remember the relief that he was dead and couldn't suffer any more, or was it that I didn't have to watch him suffer? I never talked about him.

More recently I have thought about the cruelties of our

culture, that demands a 'stiff upper lip' and no embarrassing show of emotion. I think I may have learnt how to say good–bye, and that through sorrow and conscious weeping the horror of pain and disease–riddled bodies can be allowed to sink into the background and mourning to become more joyful. When this happens, memories can be of the happiness known of a loving mother, warm and alive, of a husband walking in the sunshine across the paddock, his pipe in his mouth, his dog by his side, coming in to tea.

This is not only a report, it is also a plea for greater openness about death. Fortunately the careful work on bereavement which has been done very recently makes many things less obscure and less lonely for the dying. This coherent knowledge can amplify traditional wisdom so that consistent and thoughtful ways of care can be learnt by any ordinary intelligent person. This systematic work has certainly now begun, Lindemann was the great pioneer in this field thirty years ago (Lindemann, 1979). Parkes (1972) is, I think, a classic work, moving and straight–forward, it is essential reading for which this brief summary is no substitute.

Parkes summarizes the evidence which makes it plain that bereavement is a state of acute stress with all the features which were outlined early in this book. Cerebral disrhythmia, cardio-vascular changes, hyperactivity, panic and flight–fight reactions all occur. Being a state of distress means that it is a *normal illness*, and should be seen as such. For, not only do mental disturbances arise very commonly indeed, but individuals are prone to physical disease as well. This to the extent that the death rate among certain groups of bereaved people is very significantly higher than that of similar non–bereaved people. After the death of a close loved one, a person is at risk.

Following on from this, it is fairly clear that a sequence can be detected in the months after a death. From the storm of painfulness that goes with mourning, Parkes suggests that the following movements can be seen as the essence of it. Acute grief in the few days after a death is when distress is at its height and is best epitomized by the term *alarm*. The whole organism is in a biological state of alarm, the person is not likely to be noisy, on the contrary he is usually very quiet. But the signs of alarm are omnipresent, sleep is disturbed, there is often subdued hyper-

activity and hallucination in otherwise normal people. At this time
it is still difficult for a grieving person to believe that the loved one
has gone; he is automatically spoken to, heard around the corner,
his place even automatically laid at table. As the loss begins to
sink home a phase of *searching*, usually quite unconsciously and
automatically driven, sets in. Then this is felt as hopeless and
waves of weeping and despair at loss well up. Somehow, arising
out of these tears, periods of calmness begin to emerge which
come in *mitigation* of the loss. A person begins to warm again
inside and become interested in the world around, sad at the loss
but feeling glad to be alive.

This is laid out very neatly here, but it does not happen in this
way, for moods come and go. Past life together is relived and
re–sorted. Waves of guilt assail the bereaved person and so does
rage at being let down. Old recriminations gnaw and further guilt
wells up.

It is not only a state of distress, it is an acute depressive crisis (as
mid–life was a more chronic depressive crisis). It is a crisis striking
a person from the outside rather than arising from within. And,
like any crisis, it entails breaking of old patterns, regression,
withdrawal and then the slow discovery and testing out of a
partially new identity.

It is his shattered distress that makes it important for a mourner
to have the opportunity for conversation with another understanding person. Probably this allows a person to continue,
unconsciously, connecting those primitive patterns of earliest
conversations he had in life which we saw as part of our innate
propensity, and which soothe even an infant's distress. Sometimes
it hardly matters what is actually talked about, the tone of the
conversation itself is enough. Without this company the 'going to
bits' caused by distress and its possible hallucinations can be
terrifying. The frantic defensive manoeuvres that arise often
become wild and even self–destructive. Open destructiveness of
others is not much reported upon but it does happen. For
instance, an observer at the time reported that Shaka Zulu's grief
caused the massacre of about 7,000 people on the day after his
mother's death. Vendettas and vengeance are not openly practised
in our society but are present in hidden forms. Relatives can
become particularly unpleasant with each other, not usually at the
time of the funeral but in the months after.

271

So not only is another person a soothing presence, he can also act as a check to the mourner's ideas and more violent impulses. He can be a continuing reminder that there is enjoyable benevolence in reality which often seems empty. As well as this, being regressed, a person often needs a vigilant helper to carry out essential tasks that the mourner cannot or forgets to do.

However, most bereaved people are adults and have their own dignity. They need to be alone frequently and also to prove their own competence to themselves. Fussing is particularly seductive at this time but fundamentally choking and usually extremely irritating.

In the later phases, working out what to do next, and finding a new integrity or identity is mostly done alone. But people nearly always need to voice their ideas to check them with another person.

Remember too that, when a person takes a *long time* dying the sequence of mourning may be very different from a sudden death. In the former case much of the mourning by relatives and friends may well be done *before death* during the terminal phase. It is not uncommon in such instances, of people getting to the point of feeling, 'Oh! I wish they would hurry up and die now'. Charles II is reported as apologizing on his death bed to his impatient brother for taking an 'unconscionable time adying'.

With our greater awareness of mourning, it is not only specialists who are alerted to its dangers and its pathology, which can continue for years. The clergy, particularly aware of grief for centuries, now are tending to take special care, and other organizations like Cruze Clubs and the Samaritans are perhaps forging a new aspect of our culture.

Grief in Childhood

A young child's inner representations of his world and loved people are less stable than an adult's. He cannot explicitly recognize or manifest his sense of loss. He rarely weeps at a loved person's death. His mental organization is not sufficiently coherent for him fully to recognize the extent of his loss. This seems to raise the question as to whether little children who have lost a parent should be told about it or not and, if so, how. Very many people avoid the issue by saying they are too young to understand. And the young child, usually in a state of cataclysmic shock and confusion from the depression of others in the family, rarely asks questions about it spontaneously. He has not developed the integrative capacity to put his ideas into words. From his silence, adults then often make the assumption that grief means nothing to him.

Some people will go to remarkable lengths to prevent a child knowing about a parent's death. For example:

The mother of three young children died after a long illness. Their father decided it was best to pretend that their mother was still in hospital and would be home again one day. He asked the children's head–teacher to keep the news from them. The teacher, trying to be considerate and helpful, asked all the other children in their classes and their parents not to talk about the death. However, the school was large and many children in other classes knew what had happened and naturally talked about it. Throughout the neighbourhood, children were baffled and ill at ease, some having been told not to talk about it, but not knowing why, while others felt guilty for talking quite freely, having no instructions to the contrary. Eventually matters were sorted out by someone mentioning the problem to the teacher, who got someone to talk things over with the father.

Here the father cannot really be condemned; he was distraught after the death of his wife with three young children to look after. The teacher, too, was trying to help him. But a lie arose which caused more trouble to his children than it was worth, for not only were they surrounded by grief at home but also by embarrassment, secretiveness and confusion among their school friends.

We can see a clear and definite conclusion arising from this: that not to tell a child about the death of a parent is to withhold the truth about an important reality. Naturally enough, this does not matter if the child is too young to attach meaning to the spoken word. It would, for instance, be absurd to talk to a young baby of his parent's death. So the question arises first as to how much children understand about what has happened and secondly how much of verbal explanation is helpful. Anthony's (1971) work helps us to work out guidelines.

It seems that children dimly realise at a very early age that something terrible has happened. What is more, even if they do not *appear* to feel much at the time, the death of a parent soon becomes evident, for they never see him again. This is the main meaning of a death to a young child. A parent has disappeared and never returns. As we saw in earlier chapters, loss and separation can have very far–reaching, even life–long effects upon the child's capacity to feel and think.

The situation is made more complex by the fact that a young child cannot yet communicate his grief coherently. He may *act* in a disturbed way, but he has not yet organized his ideas in order to speak or openly weep as an older mourner does. A helping adult cannot wait until the child is openly ready to talk, because he may never be able to verbalize his grief coherently. It seems that it is best for adults to allow children to participate in the family's grief, such as in the funeral and to give explanations which are within the comprehension of the child, and then to leave it to the child to ask what questions he can as time goes by. Answers can then be simple and truthful, confined to what is understandable to the child. Over and above this, the adult must expect anxiety and disturbed behaviour from a little child about ideas and feelings which he cannot yet put into words. Sensitive people find their own ways intuitively about how to respond in these circum-stances. Specific rules and procedures are out of place. Without understanding, the child is left with his own grief in an alien, uncommunicative world. This aberration must remain if his adults do not recognize and feel his anxiety and loss, however kind they may be in other ways.

Completion of Mourning

The stage of acute distress may last for only a few weeks, but critical changes inside the individual go on less dramatically for many months and more quietly for years.

Slowly the grieving person finds a new distribution for his energies, and new objects to love and hate. And yet, if mourning is successful, awareness of the dead person is not lost. The immediate experience of the person as dead loses force and gives way to living memory of the person as he was when alive. The

274

mourner in doing this has not necessarily 'become' the dead person, though this sometimes happens, but rather he has internalized him, and has his experience of him from the past ready to use in present and future situations.

This is the natural end product of mourning. The mourners slowly come alive again. The dead person has gained a certain immortality in the memory of his loved ones. They themselves are then perhaps ready to carry on living, and are a little more prepared to die themselves one day, for other generations to carry them on also into the future.

FURTHER READING

1 Hinton, J. (1967), *Dying* (Harmondsworth: Penguin). A classic, short, must be read.
2 Kavanagh, R. E. (1972), *Facing Death* (Harmondsworth: Penguin).
3 Kubler–Ross, E. (1975), *Death, the Final Stage of Growth* (Englewood Cliffs, NJ and London: Prentice–Hall). An easily readable book by one of the world's best known and respected authorities on dying.
4 Parkes, C. M. (1972), *Bereavement* (Harmondsworth: Penguin). A classic.
5 Schneideman, E. S. (ed.) (1984), *Death, Current Perspectives* (Palo Alto, Cal.: Mayfield). This rather boring title hides a great collection of essays. A beautiful book to have by you to dip into.

Conclusion

We have come to the end of this intellectual excursion using the developmental point of view.

I hope you may find that this way of thinking will be enjoyable and useful in the future.

We can employ it when reflecting about ourselves. It is often deeply pleasant, as well as helpful, to ponder upon the more distant, as well as immediate, problems that we face at any time. We can experience that some sides of such questions arise from outside us, while others are self generated or come from within. We can often note our minds fleeing or switching off from some of the questions, usually the unpleasant or embarrassing ones. But the very exercise of meditation about present and future, however informal it is, can give us the delight of unexpected new ideas as well as checking absurdities.

We can go further and decide to *explore* ourselves in relation to each new problem that faces us. In doing this we make the avowed aim to *move the position* of our point of view away from the one we started with when the problem first came into our mind. We are in a state of readiness to *argue* within ourselves, that is to try and *combine* the understanding from both old and new points of view. It is somehow by this means, I think, that profoundly enjoyable and useful *insights* are gained in problem solving.

But it is not only with ourselves that the developmental and problem solving framework is useful. When approached by friends with a problem or by patients and clients it is surely always worthwhile to listen long and carefully to what they have to say and then to ask oneself about the setting of their life situation. We can formulate and check over, to ourselves, about such aspects as: the network of their relatives and friends, the nature of their contributions to others, as well as their special abilities and weaknesses. We can then go on roughly listing their immediate and more distant future problems that have to be faced and solved.

Within this structure we can estimate the nature and extent of their successes, competence, failures, disturbance and handicap.

We can know them better and hence perhaps be helpful as we suffer through their predicaments with them.

Suggested Films and Videos

Below is a list of suitable films for use with the chapters of this book. Each entry gives: title of film (country of origin), short description, running time in minutes, film (F) or video (V).

Unless otherwise stated all are obtainable from:

Concord Films Council Ltd.
201, Felixstowe Road,
Ipswich, Suffolk. IP3 9BJ
Tel: 0473 76012 & 715754

Chapter 1 Introduction

1 *Development* (USA) Current research methods and ideas in the study of development. — 33 F

2 *The Enchanted Loom* (UK) Account of recent discoveries about the human brain. — 50 F + V

3 *Sensory World* (USA) An examination of the mechanics of man's perception of his environment. — 33 F

Chapter 2 Pregnancy

1 *Barnet – The Child* (Sweden) Complete account of conception, gestation and delivery of a child. — 48 F

2 *The Family of Man: Birth* (UK) Part of a BBC series comparing different cultures in childbirth. — 50 F

3 *Family Matters* (UK) Part 1: Account of foetal growth, a voice over of mother's story in pregnancy. — 20 F & V
Part 2: The birth of a baby. — 20 F & V
Part 3: A home confinement. — 20 F & V

4 *Five Women, Five Births* (USA) Five births filmed together with conversations with the women. — 29 F & V

5 *Having a Baby* (UK) A series of five short films illustrating different aspects of pregnancy and childbirth. — 7-11 F & V each

6 *On the Way* (UK) First year methods of life series: follows a couple through pregnancy. — 25 F

7 *An Unremarkable Birth* (Canada) A critical discussion and record of various modern obstetric methods. — 52 F

Chapter 3 The First Months

1 *All Yours* (UK) First years of life series comparison of two two–month–old boys. — 25 F

2 *Amazing Newborn* (USA) A condensation from hours of observation of babies from one to seven days old. — 25 F

3	*Breast Feeding* (UK) Comparison of breast and bottle feeding.	30 F
4	*Breast Feeding a Special Closeness* (USA) Emotional and practical issues of breast feeding.	23 F
5	*Family Matters* Parts 3, 4, 5 and 6 (UK) Granada film about the events of the first two years.	20 F & V each
6	*Growth Through Play* (UK) A baby's awareness of environment from two to sixteen weeks.	21 F
7	*Mother Love* (USA) Social relations between mother and child starting with feeding.	20 F
8	*Mothers Own* (UK) Various women describe breast feeding.	20 F
9	*Newborn – Birthright* (Canada) Shows a baby's remarkable capacity to control his environment right from birth.	27 F
10	*Sunday's Child. The Development of an Individual* (UK) In eight parts. This is the most detailed film yet made of one infant's development over two years. Invaluable, especially for those who cannot observe an infant for themselves. Also a summary film of 45 minutes and excerpts on special topics: play, role of father, language, day minding. Hire orders to: Lynn Barnett, Iddesleigh House Clinic, 97 Heavitree Road, Exeter, Devon, EX1 2ND, Tel. 0392-76348.	60 V each part
11	*A Touch of Sensitivity* (UK) A BBC film about the importance of touching, not just about babies.	50 F

Chapter 4   The Second Six Months

1	*A Baby is Weaned* Parts 1 and 2 (UK) Two short videos to stimulate discussion about weaning and development.	7-9 V
2	*The Child Watchers* (USA) This illustrates some of Piaget's observations about development.	30 F
3	*The First Five Years: Jason and Eloise* (UK) Part 1 A five-part BBC series studies two children from birth to five years old.	30 F
4	*From Hand to Mouth* (UK) Beginning to explore.	20 F
5	*Katie's First Year* (UK) Traces the first year of one baby and her family.	25 F & V
6	*Moving Off* (UK) Mastering locomotion.	20 F
7	*Silent Speech* (UK) This BBC film illustrates Professor Montagher's observations of the correlation between children's behaviour and mothers' attitudes.	43 F & V
8	*Sunday's Child.* For details see entry under *The First Six Months*.	60 V each part

9   *A Two Year Old Goes to Hospital* (UK)   The   30 F & V
Robertsons classic about separation. Not about a
young infant but it illustrates the problem that has
been introduced in this chapter.

10   *Young Children in Brief Separation* (UK)   A series   33-43 F & V
of five complementary films by the Robertsons on
young children separated from their mothers. They
show the factors that affect the child's ability to
cope. Not about infancy but the film is included here
because separation is raised in this chapter.

Chapter 5   One Year Old

1   *The First Five Years: Jason and Eloise* (UK)   Part 1.   30 F
BBC five-part study of two children from birth to
five.

2   *Language and Development* (USA)   Recent studies   20 F
of language acquisition.

3   *Looking at Children*, Series 3, Part 1 (UK)   A study   30 F
of one–year–olds and their movements.

4   *Out of the Mouths of Babes* (Canada)   Investigates   27 F
research on the innate basis of language.

5   *Sunday's Child*. For details see entry under *The First*   60 V
Six Months.   each part

Chapter 6   Two to Three Years Old

1   *At Least Let Me Play* (UK)   Shows the importance   20 F
of play and the condition of children deprived of it.

2   *Childhood, Right of Every Child* (UK)   A group of   30 F
young children having early experiences of play.

3   *Children and Play Series* (UK)   A series of short   5 F
films about play materials and places, clay, dough,
paint, sand, water, indoors, etc.

4   *The Family of Man, Children* (UK)   BBC cross–   50 F
cultural comparison of how children are cared for.

5   *The First Five Years: Jason and Eloise* (UK)   Part 3.   30 F
Study of two children from birth to five.

6   *Looking at Children*, Series 3, Part 2 (UK)   A˙   30 F
continuation of Part 1 of this series showing the
children one year later.

Chapter 7   Three to Five Years

1   *The First Five Years: Jason and Eloise* (UK)   Parts   30 F
4 and 5. A longitudinal study of two children from
birth to five.

2   *Looking at Children*, Series 3, Part 3 (UK)   The   20 F
same children seen earlier in the series now three
years old.

3 *Playing* (UK) This looks at a cross section of play   28 F
invented by children of four to eleven.

Chapter 8   Early School

1 *Buckets, Spades and Hand Grenades* (UK) TV film   52 F
of observation of children in playground from
different backgrounds.

2 *Discovery and Experience*, 10 parts (UK) BBC   30 F
series showing different aspects of education in   each
primary schools.

3 *Four Teachers* (Canada) Teaching methods in   60 F
Canada, Japan, Poland and Puerto Rico compared.

4 *Seven Up* (UK) Classic Granada study of seven–   30 F & V
year–old children from different backgrounds.
Followed by 'Seven Plus Seven', 'Twenty–One' and
'Twenty–Eight' all at seven year intervals.

Chapter 9

1 *The Family of Man. Teenagers* (UK) BBC cross–   50 F
cultural comparisons.

2 *The Searching Years* (USA) Part 1–6. A series to   12 F
open discussion on issues of adolescence.   each

3 *The Schools* (UK) BBC study of comprehensives in   75 F
Britain.

4 *Seven Plus Seven* (UK) The first sequel to *Seven Up*.   60 F & V

5 *The Space Between Words* (UK) Family   75 F
School   55 F
BBC films about communications and their break-
down.

Chapters 10, 11, 12, 13 and 14

For these years of adulthood there are a vast array of films about specific
topics and questions but few that are both general and searching. For
titles consult a catalogue such as that of Concord Films.

Chapter 15   Old Age

1 *The Family of Man: Old Age* (UK(Cross–cultural   50 F
comparison. BBC series.

2 *Forward to Retirement* 1 Health in Later Years.   30 F
2 Woman's View.   30
Stresses importance of keeping active and interested.

3 *I Don't Want to be a Burden* (UK) The world seen   50 F & V
through the eyes of very old people.

4 *Three Grandmothers* (Canada) The lives of grand-   30 F
mothers in Nigeria, Canada and Brazil.

5 *What Shall We Do With Granny?* (UK) Parts 1 and   50 F & V
2. BBC film looking at how different families have   each

281

tried to resolve the problem of how to look after their old people.

Chapter 16 Dying, Grief and Mourning

1 *Begin with Goodbye* (USA) Seven parts. A series of films to make us think about changes in ordinary life. Particularly concerned with the losses when going through developmental crises. 'Time to Cry' and the 'Death of Ivan Ilych' deal with death and grief specifically. 28 F each

2 *Facing Death* (UK) A series of films for TV about people dying and their bereaved relatives. 30 F & V each

3 *Grief* (UK) Interviews and discussion of ordinary grief by bereavement. 52 F & V

4 *Hospice – St Christophers* (UK) An explanation of the method of this world-known hospice. 38 F & V

References

The starred references are specialized technical sources. They are of use only for detailed further study.

*Abraham, K. (1927), *Selected Papers on Psychoanalysis* (London: Hogarth).
*Ainsworth, M. and Belhar, M. L. (1978), *Patterns of Attachment* (Hillside, New York: Lawrence Erlbaum).
Aitchison, J. (1983), *The Articulate Mammal* (London: Hutchinson).
Anthony, E. J. and Benedek, T. (eds) (1970), *Parenthood* (Boston: Little Brown).
Anthony, S. (1971), *The Discovery of Death in Childhood and After* (Harmondsworth: Penguin).
Archer, J. and Lloyd, B. (1982), *Sex and Gender* (Harmondsworth: Penguin).
Arendt, H. (1969), *The Human Condition* (Chicago: University of Chicago Press).
Argyle, M. (1975), *Bodily Communication* (London: Methuen).
Asher, J. (1984), *Silent Nights* (London: Pelham).
Backett, K. C. (1982), *Mothers and Fathers* (London: Macmillan).
Bailey, C. (1984), *A Loving Conspiracy* (London: Quartet).
Beadle, M. (1970), *A Child's Mind* (London: Methuen).
Becker, E. (1971), *The Birth and Death of Meaning* (Harmondsworth: Penguin).
Beckett, B. S. (1981), *Illustrated Human Biology* (Oxford: OUP).
Bell, E. W. and Vogel, N. (1968), *A Modern Introduction to the Family* (London: Macmillan).
*Berlinsky, E. B. and Biller, H. B. (1982), *Parental Death and Psychological Development* (Lexington, Mass.: Lexington Books).
Bernard, J. (1972), *The Future of Marriage* (Harmondsworth: Penguin).
Berne, E. (1964), *Games People Play* (Harmondsworth: Penguin).
*Biller, H. B. (1974), *Paternal Deprivation* (Lexington, Mass.: Lexington Books).
Black, C. (1983), *Married Women Work* (London: Virago).
*Blos, P. (1962), *On Adolescence* (London: Macmillan).
*Blos, P. (1979), *The Adolescent Passage* (New York: IUP).
Bluebond–Langer, M. (1978), *The Private Worlds of Dying Children* (Princeton, NJ: Princeton University Press).
Boden, M. (1979), *Piaget* (London: Fontana).
*Bok, S. (1978), *Lying, Moral Choice in Public and Private Life* (New York: Pantheon).
Bower, T. (1977), *A Primer of Infant Development* (Reading and San Francisco: W. H. Freeman).

Bower, T. (1982), *Development in Infancy* (Reading and San Francisco: W. H. Freeman).

Bowlby, J. (1953), *Child Care and the Growth of Love* (Harmondsworth: Penguin).

Bowlby, J. (1969), *Attachment* (Harmondsworth: Penguin).

Bowlby, J. (1973), *Separation* (Harmondsworth: Penguin).

Bowlby, J. (1980), *Loss* (Harmondsworth: Penguin).

Boyle, D. G. (1969), *Students Guide to Piaget* (Oxford: Pergamon).

*Breen, D. (1975), *The Birth of a First Child* (London: Tavistock).

Britton, J. (1985), *Language and Learning* (Harmondsworth: Penguin).

Bromley, D. B. (1974), *The Psychology of Human Ageing* (Harmondsworth: Penguin).

Brown, D. and Pedder, J. (1979), *Introduction to Psychotherapy* (London: Tavistock).

Brubaker, T. H. (1983), *Family Relationships in Later Life* (Beverley Hills: Saga Press).

Bruner, J. (ed.) (1976), *Play, Its Role in Development and Evolution* (Harmondsworth: Penguin).

Bryant, P. (1974), *Perception and Understanding in Young Children* (London: Methuen).

*Burstock, R. H. and Shana, E. (eds) (1976), *Handbook of Aging and the Social Sciences* (New York: Van Nostrand Reinhold).

Butcher, H. J. (1968), *Human Intelligence, Its Nature and Assessment* (London: Methuen).

*Cameron, N. (1963), *Personality Development and Psychopathology* (Boston: Houghton Mifflin).

Caplan, G. (1964), *Principles of Preventive Psychiatry* (London: Tavistock).

Carter, E. A. and McGoldrick, M. (1980), *The Family Life Cycle* (New York: Gardener Press).

Carter, M. (1966), *Into Work* (Harmondsworth: Penguin).

Carter, N. (1980), *Development, Growth and Ageing* (London: Croom Helm).

Casement, P. (1985), *On Learning from the Patient* (London: Tavistock).

Cass, J. E. (1977), *The Significance of Children's Play* (London: Batsford).

Cath, S. F., Gurwitt, A. R. and Ross, J. M. (1982), *Father and Child* (Boston: Little Brown).

Chamberlain, G. (1969), *The Safety of the Unborn Child* (Harmondsworth: Penguin).

*Chassequet–Smirgel, J. *et al.* (1980), *Female Sexuality* (Ann Arbor, MI: University of Michigan Press).

Chodorow, N. (1978), *The Reproduction of Mothering* (Berkeley: University of California Press).

Clarke, A. M. and Clarke, A. D. (1975), *Early Experience, Myth and Evidence* (London: Open Books).

Clulow, C. F. (1982), *To Have and to Hold* (Aberdeen: Aberdeen University Press).

*Cohen, R. (ed.) (1982), *Children's Conception of Spacial Relationships* (San Francisco, CA: Jossey Bass).

Danziger, K. (1970), *Readings in Child Socialisation* (Oxford: Pergamon).

Davie, R., Butler, N. and Goldstein, H. (1972), *From Birth to Seven* (London: Methuen).

Davies, D. R. and Shackleton, V. J. (1975), *Psychology and Work* (London: Methuen).

*Decarie, T. G. (1974), *Infants Reactions to Strangers* (New York: IUP).

de Mause, L. (ed.) (1974), *The History of Childhood* (London: Souvenir Press).

*Deutsch, H. (1944), *The Psychology of Women*, Vol. II (London: Research Books).

de Villiers, J. and P. (1977), *Early Language* (London: Fontana).

Dicks, H. V. (1967), *Marital Tensions* (London: Routledge & Kegan Paul).

Dorr, D., Zax, M. and Danner, T. (eds) (1978), *The Psychology of Discipline* (New York: IUP).

Douglas, J. and Richman, N. (1984), *My Child Won't Sleep* (Harmondsworth: Penguin).

Dunn, J. (1984), *Sisters and Brothers* (London: Fontana).

*Ellenburger, H. F. (1970), *The Discovery of the Unconscious* (New York: Basic Books).

Erikson, E. H. (1963), *Childhood and Society* (London: Paladin).

Erikson, E. H. (1964), *Insight and Responsibility* (New York: Norton).

Erikson, E. H. (1968), *Identity, Youth and Crisis* (London: Faber).

*Escalona, S. (1969), *The Roots of Individuality* (London: Tavistock).

Fagin, L. and Little, M. (1984), *The Forsaken Families* (Harmondsworth: Penguin).

Fenichel, O. (1946), *The Psychoanalytic Theory of Neurosis* (London: Routledge & Kegan Paul).

Fisher, S. (1973), *Understanding the Female Orgasm* (Harmondsworth: Penguin).

Fleming, C. M. (1967), *Adolescence. Its Social Psychology* (London: Routledge & Kegan Paul).

*Floyd, A. (ed.) (1979), *Cognitive Development in the School Years* (London: Croom Helm and Open University).

Fogelman, K. (1983), *Growing Up in Great Britain* (London: Macmillan).

Fraiberg, S. H. (1959), *The Magic Years* (London: Methuen).

Fraiberg, S. H. (1977), *Every Child's Birthright. In Defence of Mothering* (New York: Basic Books).

Freud, A. (1928), *Introduction to the Technique of Child Analysis* (New York: NMD Publishing Co.).

Freud, A. (1937), *The Ego and Mechanisms of Defence* (London: Hogarth).

*Freud, S. (1900), *The Interpretation of Dreams*, Standard Edition, Vol. IV (London: Hogarth and Harmondsworth: Penguin).

Freud, S. (1905), *Three Essays on Sexuality*, Standard Edition, Vol. XII (London: Hogarth).

Freud, S. (1913), *Totem and Taboo*, Standard Edition, Vol. XII (London: Hogarth).

*Freud, S. (1915a), *Instincts and their Vicissitudes*, Standard Edition, Vol. XIV (London: Hogarth).

Freud, S. (1915b), *Introductory Lectures on Psychoanalysis*, Standard Edition, Vol. XV (London: Hogarth and Harmondsworth: Penguin).

*Freud, S. (1923), *The Ego and the Id*, Standard Edition, Vol. XXIII (London: Hogarth and Harmondsworth: Penguin).

Fromm, E. (1957), *The Art of Loving* (London: Allen & Unwin).

Fromm, E. (1973), *Anatomy of Human Destructiveness* (Harmondsworth: Penguin).

Furman, E. (1974), *A Child's Parent Dies* (New Haven Conn. and London: Yale University Press).

Garvey, C. (1977), *Play* (London: Fontana).

Geer, J., Heinman, J. and Lentenberg, H. (1984), *Human Sexuality* (Englewood Cliffs, NJ: Prentice–Hall).

Giele, J. Z. (1982), *Women in the Middle Years* (New York: Wiley).

Gilligan, C. (1982), *In a Different Voice* (Cambridge, Mass.: Harvard University Press).

Gorer, G. (1965), *Death, Grief and Mourning in Contemporary Britain* (London: Cresset).

Gorer, G. (1971), *Sex and Marriage in England Today* (London: Nelson).

Goffman, I. (1959), *The Presentation of Self in Everyday Life* (Harmondsworth: Penguin).

Green, L. (1968), *Parents & Teachers, Partners or Rivals* (London: Allen & Unwin).

Green, M. (1978), *Goodbye Father* (London: Routledge & Kegan Paul).

Grosskopf, D. (1983), *Sex and the Married Woman* (London: Columbus).

Hareven, T. K. and Adams, K. J. (eds) (1982), *Ageing in Life Course Transitions* (London: Tavistock).

Hayley, J. (1968), *The Family of the Schizophrenic* in Handel, G. (ed.) *The Psychosocial Interior of the Family* (London: Allen & Unwin).

Henry, J. (1971), *Pathways to Madness* (New York: Vintage).

Hinton, J. (1967), *Dying* (Harmondsworth: Penguin).

Holt, J. (1982), *How Children Fail* (Harmondsworth: Penguin).

Holt, J. (1983), *How Children Learn* (Harmondsworth: Penguin).

Hunt, M. (1982), *The Universe Within* (Brighton: Harvester).

Hunt, S. and Hilton, J. (1975), *Individual Development and Social Experience* (London: Allen & Unwin).

Hutt, C. (1975), *Sex Differences in Behaviour* (London: Crosby Lockwood).

Inglis, K. (1982), *Must Divorce Hurt Children?* (London: Temple Smith).

Isaacs, S. (1930), *Intellectual Growth in Young Children* (London: Routledge & Kegan Paul).

Isaacs, S. (1933), *Social Development in Young Children* (London: Routledge & Kegan Paul).

Jaques, E. (1970), *Work Creativity and Social Justice* (New York: IUP).

Kahn, J. and Wright, S. E. (1980), *Human Growth and the Development of Personality* (Oxford: Pergamon).

Kavanagh, R. E. (1972), *Facing Death* (Harmondsworth: Penguin).

Kelly, G. A. (1963), *A Theory of Personality* (New York: Norton).

Kelmer Pringle, M., Butler, N. R. and Davie, R. (1966), *1100 Seven Year Olds* (London: Longman).

Kelmer Pringle, M. (1974), *The Needs of Children* (London: Hutchinson).

*Khan M. M. (1974), *The Privacy of the Self* (London: Hogarth).

Kemiston, K. (1968), *Young Radicals* (New York: Harcourt Brace).

Kitzinger, S. (1979), *The Experience of Breastfeeding* (Harmondsworth: Penguin).

Kitzinger, S. (1980), *Pregnancy and Childbirth* (London: Michael Joseph).

Kitzinger, S. (1984), *The Experience of Childbirth* (Harmondsworth: Penguin).

*Klaus, M. and Kennett, J. (1976), *Maternal–Infant Bonding* (St. Louis: C. V. Mosby).

Klein, J. (1965), *Samples of English Culture* (London: Routledge & Kegan Paul).

*Klein, M. (1932), *The Psychoanalysis of Children* (London: Hogarth).

*Klein, M. (1948), *Contributions to Psychoanalysis* (London: Hogarth).

*Klein, M. (1952), *Developments in Psychoanalysis* (London: Hogarth).

Kline, P. (1972), *Fact and Fantasy in Freudian Theory*, (London: Methuen).

Kohlberg, L. (1969), *State & Sequence* in Goslin, D. A. (ed.) *Handbook of Socialisation Theory* (Chicago: Rand McNally).

*Kohut, H. (1971), *The Analysis of the Self* (London: Hogarth).

Konner, M. (1982), *The Tangled Webb, Biological Constraints on the Human Spirit* (Harmondsworth: Penguin).

*Krames, L., Pliner, P. and Alloway, T. (1975), Vol. I Non Verbal Communication Vol. II Non Verbal Communication, Aggression Vol. III Non Verbal Communication, Attachment (New York: Plenum).

Kubler–Ross, E. (1972), *Questions and Answers on Death and Dying* (London and New York: Macmillan).

Kubler–Ross, E. (1975), *Death, the Final Stage of Growth*. (Englewood Cliffs NJ: and London: Prentice–Hall).

Laing, K. D. and Esterson, A. (1964), *Sanity, Madness and the Family* (Harmondsworth: Penguin).

Lamb, M. E. (ed.) (1981), *The Role of the Father in Child Development* (New York: J. Wiley.)

Langer, M. (1957), *Learning to Live as a Widow* (New York: Gilbert Press).

Laufer, M. (1974), *Adolescent Disturbance and Breakdown* (Harmondsworth: Penguin).

*Laufer, M. and M. E. (1984), *Adolescence and Developmental Breakdown* (Princeton and London: Yale University Press).

Leach, P. (1974), *Babyhood* (Harmondsworth: Penguin).
Leach, P. (1983), *A–Z Parents Guide* (London: Allen Lane).
Lee, V. (1979), *Language Development* (London: Croom Helm).
*Levi–Strauss, L. (1966), *The Savage Mind* (Chicago: University of Chicago Press).
Lewin, R. (ed.) (1975), *Child Alive* (London: Temple Smith).
*Lewis, R. and Brooks, J. (1979), *Sexuality, Cognition and Acquisition of the Self* (New York: Plenum).
*Lichtenstein, H. (1977), *The Dilemma of Human Identity* (New York: J. Aronson).
Lidz, T. (1963), *The Family and Human Adaptation* (London: Hogarth).
Lindemann, E. and E. (1979), *Grief, Studies in Crisis Intervention* (New York: J. Aronson).
Lynd, H. M. (1958), *Shame and the Search for Identity* (London: Routledge & Kegan Paul).
Maccoby, E. C. and Jacklin, C. N. (1975), *The Psychology of Sex Differences* (Oxford: OUP).
MacFarlane, A. (1977), *The Psychology of Childbirth* (London: Fontana).
*Mahler, M. S., Pine, F. and Bergman, A. (1975), *The Psychological Birth of the Human Infant* (London: Hutchinson).
*Masters, W. M. and Johnson, V. E. (1966), *The Human Sexual Response* (New York: Bantam).
*Matte Blanco, I. (1975), *The Unconscious as Infinite Sets* (London: Duckworth).
Matteson, D. R. (1975), *Adolescence Today* (Holmwood Ill.: Dorsey Press).
May, R. (1967), *Psychology and the Human Dilemma* (Princeton: Van Nostrand).
May, R. (1980), *Sex and Fantasy* (New York: W. W. Norton).
Mayerson, S. (ed.) (1975), *Adolescence and the Crises Adjustment* (London: Allen & Unwin).
McKee, L. and O'Brien, M. (1982), *The Father Figure* (London: Tavistock).
McNeill Taylor, L. (1983), *Living with Loss* (London: Fontana).
*Meade, G. H. (1932), *Mind, Self and Society* (Chicago, University of Chicago Press).
Millar, S. (1968), *The Psychology of Play* (Harmondsworth: Penguin).
Miller, J. (1983), *The Human Body* (London: J. Cape).
Miller, J. B. (1976), *Towards a New Psychology of Women* (Harmondsworth: Penguin).
Milner, M. (1957), *On Not Being Able to Paint* (London: Heinemann).
Montessori, M. and Carter, B. B. (1983), *The Secret of Childhood* (London: Sangam).
Morrison, H. L. (1983), *Children of Depressed Parents* (New York: Grune & Stratton).
*Neugarten, B. C. (1968), *Middle Age and Ageing* (Chicago: University of Chicago Press).

REFERENCES

Newsom, J. and E. (1963), *Patterns of Infant Care* (Harmondsworth: Penguin).

Newsom, J. and E. (1969), *Seven years old in a Home Environment* (Harmondsworth: Penguin).

Nichols, J. (1975), *Men's Liberation* (Harmondsworth: Penguin).

Norman, W. H. and Scaramella, W. (1980), *Mid–Life* (New York: Brunner Mazel).

Oakley, A. (1974) (a), *Housewife* (Harmondsworth: Penguin).

Oakley, A. (1974) (b), *Sociology of Housework* (Oxford: Martin Robertson).

Oakley, A. (1975), *Birth Without Violence* (London: Wildwood House).

Oakley, A. and Mitchell, J. (1978), *The Rights and Wrongs of Women* (Harmondsworth: Penguin).

Oakley, A. (1979), *Becoming a Mother* (Oxford: Martin Robertson).

Oakley, A. (1980), *Women Confined* (Oxford: Martin Robertson).

Oakley, A. (1981), *Subject Women* (London: Fontana).

Olson, D. H. *et al.* (1983), *Families. What Makes them Work* (Beverley Hills: Sage).

Ousted, C. and Taylor, D. C. (eds.) (1972), *Gender Differences* (Edinburgh: Churchill Livingstone).

Overton, W. F. and Gallagher, J. M. (1977), *Knowledge & Development* (London and New York: Plenum).

*Palombo, S. R. (1978), *Dreaming and Memory* (New York: Basic Books).

Parke, R. D. (1981), *Fathering* (London: Fontana).

Parker, G. (1983), *Parental Overprotection* (New York: Grune & Stratton).

Parkes, C. M. (1972), *Bereavement* (Harmondsworth: Penguin).

Parker, C. M. and Weiss, R. S. (1983), *Recovery from Bereavement* (New York: Basic Books).

Pepper, L. G. and Knapp, R. J. (1980), *Motherhood and Mourning* (New York: Praeger).

Persig, R. (1974), *Zen and the Art of Motorcycle Maintenance* (London: Corgi).

*Piaget, J. (1929), *The Child's Concept of the World* (London: Paladin).

*Piaget, J. (1935), *Moral Judgements of the Child* (London: Routledge & Kegan Paul).

*Piaget, J. (1950), *The Psychology of Intelligence* (London: Routledge & Kegan Paul).

*Piaget, J. (1951), *Play Dreams and Imitation in Childhood* (London: Routledge & Kegan Paul).

*Piaget, J. (1953). *The Origins of Intelligence in the Child* (London: Routledge & Kegan Paul).

*Piaget, J. (1955), *The Child's Construction of Reality* (London: Routledge & Kegan Paul).

*Piaget, J. (1970), *The Child's Conception of Movement and Speed* (London: Routledge & Kegan Paul).

Piaget, J. (1972), *The Child and Reality* (Harmondsworth: Penguin).

Phillips, A. and Rakusen, J. (eds.) (1978), *Our Bodies Ourselves* (Original American edition by Boston Women's Health Books Collective). (Harmondsworth: Penguin).

*Phillips, J. L. (1975), *The Origins of Intellect* (Reading and San Francisco CA: W. H. Freeman).

Rapoport, R. and R. and Strelitz, Z. (1979), *Fathers, Mothers and Others* (London: Routledge & Kegan Paul).

*Rayner, E. (1981), Infinite Experiences. p. 463 *Int. Journal of Psychoanalysis* Vol. 62 (London: Balliere Tindall).

Reiss, D. (1981), *The Family's Construction of Reality* (Cambridge Mass: Harvard University Press).

Richards, M. (1980), *World of the Newborn* (New York: Harper Row).

Roberts, M. and Tamburri, J. (1981), *Child Development 0–5* (London: Holmes McDougall).

Rochlin, G. (1973), *Man's Aggression: the Defence of the Self* (New York: IOP).

Roland, P. (1973), *Children Apart* (New York: Pantheon).

*Rosen, I. (1979), *Sexual Deviation* (Oxford: OUP).

Rosen, C. H. (1973), *The Language of Primary School Children* (Harmondsworth: Penguin).

Ruitenbeck, H. M. (1973), *Homosexuality* (London: Souvenir).

Rutter, M. (1972), *Maternal Deprivation Reassessed* (Harmondsworth: Penguin).

Rutter, M. and Maugham, B. (1981), *15000 Hours* (Shepton Mallett: Open Books).

Rycroft, C. (1968), *A Critical Dictionary of Psychoanalysis* (London: Nelson).

Saunders, C. (1959), *Care of the Dying* (London and New York: Macmillan).

Schaffer, K. (1977), *Mothering* (London: Fontana).

Scharff, D. E. (1982), *The Sexual Relationship* (London: Routledge & Kegan Paul).

Schneideman, E. S. (ed.) (1984), *Death Current Perspectives* (London: Routledge & Kegan Paul). (Palo Alto, Cal.: Mayfield).

Schofield, M. (1965), *The Sexual Behaviour of Young People* (Harmondsworth: Penguin).

Schofield, M. (1976), *Promiscuity* (London: Gollancz).

Schofield, M. (1973), *Sexual Behaviour of Young Adults* (Harmondsworth: Penguin).

Segal, H. (1973), *An Introduction to the Work of Melanie Klein* (London: Hogarth).

Segal, J. (1985), *Fantasy in Everyday Life* (Harmondsworth: Penguin).

Shaffer, D. and Dunn, J. (1980), *1st Year of Life* (Chichester and New York: Wiley).

Sharpe, S. (1976), *Just Like a Girl* (Harmondsworth: Penguin).

Shayer, R. and Adey, P. (1981), *Towards a Science of Science Teaching* (London: Heinemann).

Sheridan, M. (1968), *The Developmental Progress of Infants and Young Children* (London: HMSO).

Shor, J. and Sanville, J. (1978), *Illusions in Loving* (Harmondsworth: Penguin).

Skinner, B. F. (1953), *Science and Human Behaviour* (New York: Macmillan).

*Skynner, R. S. (1976), *One Flesh Separate Persons* (London: Constable).

Skynner, R. S. and Cleese, J. (1983), *Families and How to Survive Them* (London: Methuen).

Slukin, W., Herber, M. and Slukin, A. (1983), *Maternal Bonding* (Oxford: Blackwell).

Smelser, N. and Erikson, E. (1980), *Theories of Work and Love* (London: Grant McIntyre).

Smith, A. (1970), *The Body* (Harmondsworth: Penguin).

Smith, A. (1984), *The Mind* (Harmondsworth: Penguin).

*Smith, A. J. W. and Danielson, J. (1982), *Anxiety and Defensive Strategies in Childhood and Adolescence* (New York: IUP).

Smith, P. K. (1984), *Play in Animals and Humans* (Oxford: B. Blackwell).

Snow, E. (1968), *Red Star over China* (Harmondsworth: Penguin).

*Spitz, Rene A. and Cobliner, W. Godfrey (1965), *The First Year of Life* (New York: IUP).

Stafford Clark, D. and Smith, A. (1983), *Psychiatry for Students* (London: Allen Lane).

Stern, D. (1977), *The First Relationship* (London: Fontana).

Stevens, J. (1979), *Adult Life Processes* (Palo Alto, Cal.: Mayfield).

Storr, A. (1960), *The Integrity of the Personality* (Harmondsworth: Penguin).

Sudnow, D. (1967), *Passing On* (Englewood Cliffs NJ and London, Prentice–Hall).

Sylva, K. and Hunt, I. (1979), *Child Development a First Course* (Oxford: B. Blackwell).

Tanner, I. J. (1973), *Loneliness* (New York: Harper Row).

Terkel, S. (1985), *Working* (Harmondsworth: Penguin).

Tessman, C. H. (1978), *Children of Parting Parents* (New York: J. Aronson).

Thomas, B. and Collard, J. (1979), *Who Divorces* (London: Routledge & Kegan Paul).

Tiefer, L. (1983), *Human Sexuality* (London and New York: Harper Row).

Tillich, P. (1952), *The Courage to Be* (London: Fontana).

Tough, J. (1977), *The Development of Meaning* (London: Allen & Unwin).

Townsend, P. (1963), *The Family Life of Old People* (Harmondsworth: Penguin).

Tunnadine, P. (1983), *The Making of Love* (London: Jonathan Cape).

Turiel, E. (1977), *Moral Development* (London: Fontana).

Turner, J. (1975), *Cognitive Development* (London: Methuen).

Turner, J. (1980), *Made for Life. Coping Competence and Cognition* (London: Methuen).

Verney, T. and Kelly, J. (1982), *The Secret Life of the Unborn Child* (London: Sphere).

*Vinyk, R. (1981), *Overview and Critique of Piaget's Genetic Epistomology* (London: Academic Press).

Visher, E. H. and J. S. (1979), *Step Families* (New York: Bruner Mazel).

*Wagner, D. A. and Stevenson, H. W. (1982), *Cultural Perspectives on Child Development* (San Francisco: W. H. Freeman).

Wald, E. (1981), *The Remarried* (New York: Family Service Assn.).

Walsh, F. (ed.) (1982), *Normal Family Processes* (London: New York: Guildford).

Weiss, R. S. (1975), *Marital Separation* (New York: Basic Books).

Weiss, R. S. (1979), *Going it Alone* (New York: Basic Books).

West, D. J. (1960), *Homosexuality* (Harmondsworth: Penguin).

White, R. W. (1975), *Lives in Progress* (New York: Holt Rinehart).

Willmott, P. (1966), *Adolescent Boys of East London* (Harmondsworth: Penguin).

*Winnicott, D. W. (1958), *Through Paediatrics to Psychoanalysis* (London: Hogarth).

Winnicott, (1964), *The Child, The Family and the Outside World* (Harmondsworth: Penguin).

Winnicott, D. W. (1965), *The Maturational Processes and the Facilitating Environment* (London: Hogarth).

Wolman, B. B. (ed.) (1979), *Handbook of Dreams* (New York: Van Nostrand Reinhold).

Yarrow, C. J., Rubinstein, J. L. and Pedersen, F. A. (1977), *Infant and Environment* (New York: J. Wiley).

292

Index